G. ANGELINI • D. BONAMONTE

Aquatic Dermatology

Springer-Verlag Italia Srl.

G. Angelini · D. Bonamonte

Aquatic Dermatology

Springer

Prof. GIANNI ANGELINI, M.D.
Department of Internal Medicine, Immunology and Infectious Diseases
Unit of Dermatology
Head of Dermatologic Clinic II
University of Bari, Italy

Dr. DOMENICO BONAMONTE, M.D.
Department of Internal Medicine, Immunology and Infectious Diseases
Unit of Dermatology
University of Bari, Italy

© Springer-Verlag Italia 2001
English translation by Mary V.C. Pragnell, B.A. (HONS.)

Fig. on front cover: *Hyppocampus guttulatus* (sea horse).
Fig. on back cover: Figured vesico-bullous dermatitis from *Anemonia sulcata* (the same case as
in Fig. 3.43). Reproduced with permission from: Angelini A, Vena GA (1997)
Dermatologia professionale e ambientale. Vol. I. ISED, Brescia

© Springer-Verlag Italia 2002
Originally published by Springer-Verlag Italia, Milano in 2002
ISBN 978-88-470-2161-7   ISBN 978-88-470-2095-5 (eBook)
DOI 10.1007/978-88-470-2095-5

Typesetting, printing and binding: Centro Grafico Ambrosiano, San Donato Milanese (MI)
Cover design: Simona Colombo

SPIN: 10858293

Preface

This updated and extended edition of a work first published in Italian 10 years ago owes its revised production in both Italian and English to a series of factors. Foremost among them is the ever growing number of skin diseases caused by aquatic organisms, affecting the immense population that flocks to the water for holiday, sports and professional activities. Aquatic skin diseases are no longer only a seasonal affliction but can be observed at any period, thanks to the boom of aquatic holidaymaking throughout the year.

The volume of literature in the field of aquatic dermatology is rapidly expanding. In the USA, one of the nations that is particularly attuned to the problem, a Bulletin, the "Jellyfish Sting Newsletter" has been issued six-monthly for the last 10 years. It is edited by Prof. Joseph W Burnett (Department of Dermatology, University of Maryland School of Medicine, International Consortium for Jellyfish Stings, 405 W Redwood St., 6th floor, Baltimore, Maryland 21201, USA), one of the great pioneers and researchers in the field of skin reactions to marine organisms.

The widespread passion for aquariums, filled with tropical saltwater or freshwater animals, together with various pathogenic bacteria, has also contributed to the increase in aquatic skin diseases.

In this new edition of the book, aetiological and clinical aspects have been thoroughly updated and the illustrations renewed. Greater attention has also been paid to aquatic biotic agents from tropical countries, that can now be observed everywhere as "imported" diseases.

Finally, the book aims to make a modest contribution to the knowledge of some aquatic animals that have only developed a poisonous apparatus for protective purposes. A fuller understanding of these fascinating creatures may help people to better appreciate the beauty of the aquatic environment and to enjoy this enthralling habitat at lesser risk.

Acknowledgements

This book is a revised and updated version in both Italian and English of a monograph published in Italian in 1991. Cordial thanks went in the previous edition to all the people who had contributed to it, and for much of the work collected in this volume these thanks are heartily renewed. However, various other people have contributed photographic material of superb quality to the present edition, and it is with a great sense of gratitude that we acknowledge our debt to them.

Prof. G. Marano, from the Laboratory of Marine Biology at the University of Bari, kindly provided figures 3.5, 6.3, and 10.2.

Dr. M. Capraro, from the Dermatology and Clinical Allergology Service of the Ospedali Civili Riuniti in Sciacca, produced figures 3.20 and 3.29.

Dr. Pablo Helman, co-author with Susanna Volpe of the interesting book on marine biology "Il Quaderno Blu" (Provincia di Imperia Ed., 1995), most generously contributed the following figures: cover photo (sea horse), 2.1, 3.1, 3.18, 3.19, 3.22, 3.47, 3.48, 4.19, 5.1, 6.2, 8.1, 8.2, 9.1.

The lavish contribution of the Fotoarchivio G.R.O. Sub Catania of Dr. Filippo Massari and Fabrizio Frixa includes figures 3.21, 4.1, 4.2, 4.3, 4.20, 5.2, 5.3, 5.4, 6.1, 7.1, 9.2, 9.5, 10.1, 10.5, 10.10, 10.11, 10.12, 10.13.

Dr. Raffaele Filotico took figure 11.5.

Mr. Giacomo Borraccino is the eclectic illustrator.

The contribution of our dear friends Prof. Bruno Bianchi and Mr. Mario Placido, consisting of revision and layout of the volume, has been of incalculable importance.

Finally, we are very grateful to Springer-Verlag Italia for their assiduous commitment to producing this volume.

Gianni Angelini
Domenico Bonamonte

The Authors and Springer-Verlag Italia would like to thank the Publishers ISED, Poletto, Piccin, for having kindly authorized the reproduction of the following figures and tables:

Figs. 3.3, 3.6, 3.12, 3.14, 3.15, 3.16, 3.26, 3.32, 3.34, 3.35, 3.36, 3.38, 3.39, 3.41, 4.5, 4.6, 4.9, 4.12, 4.15, 4.17, 4.18, 9.3, 9.4, 9.6, 9.9, 9.12, 10.4, 10.7, 10.8, 10.9, 11.3, 11.4, 11.8, 11.9, 11.11, 11.13, 12.1, 12.2, and the inset photo on the back cover (reproduced with permission from: *Dermatologia professionale e ambientale*, Vols. I-III, by G. Angelini and G.A.Vena. ISED, Brescia, 1997, 1999).

Figs. 3.33, 3.40, 3.42, 3.43, 4.8, 4.16, and Table 3.3 (reproduced with permission from: *Dermatologia di importazione*, by S. Veraldi and R. Caputo. Poletto Editore, Milano, 2000).

Figs. 3.45, 9.11 (reproduced with permission from: *Trattato di dermatologia*, Vol. II, by A. Giannetti. Piccin, Padova, 2001).

Tables 3.2, 8.1 (reproduced with permission from: *Gli animali marini velenosi e le loro tossine*, by F. Ghiretti and L. Cariello. Piccin, Padova, 1984).

Contents

1 Introduction

The hydrosphere

Throughout history, man has migrated to try and find places to live near water - the sea, rivers or lakes. Nowadays in particular, in our technologically advanced age offering so vast a choice of leisure activities, "water activities", even just during the short holiday period, have become a must for us all. What is more, they allow us to enjoy two other ingredients that are generally considered to be antidotes to the stresses of everyday life, in other words nudity and exposure to the sun.

In recent decades more than ever, during warm weather the coasts have been literally invaded by millions of holidaymakers attracted by the chance to practise water and underwater sports. However, although these enthusiasts are fascinated by the marine and submarine panorama, they are often ignorant of the flora and fauna that populate the aquatic environment and above all of the lurking dangers it hides.

Diseases and accidents caused by the aquatic environment have therefore increased year by year, giving rise to the rapid development of Aquatic Medicine, now a specialist field in its own right: indeed, great attention is now paid to the various professional underwater diving diseases. Instead, Aquatic Dermatology seems to have been comparatively neglected, despite the enormous number of patients with aquagenic dermatoses.

Oceans, rivers, lakes, ponds, swimming pools and aquariums all contain innumerable animal and vegetable organisms of variable sizes, including myriads of microscopic organisms. During the course of evolution, many aquatic species have developed natural defence and offence mechanisms against their natural foes, and can sting or bite, often being equipped with a venomous apparatus. Unfortunately, these self-protective mechanisms are sometimes turned against chance or involuntary aggressors, such as swimmers, underwater divers and fishermen.

Poisonous bites or stings can induce not only various dermatological pictures but also systemic reactions, often of a serious or even fatal nature. In any case, even when the damage is not serious, the very fact of being attacked in deep water can cause the swimmer to panic, and thus be a danger in itself. Most fatal accidents, that are fortunately rare, are not in fact due to the toxicity of the poison injected but to the functional impotence it triggers, which can paralyse the victim's ability to swim or to resurface correctly from underwater.

From the medical viewpoint, the complex problems linked to poisonous marine organisms are still far from having been completely elucidated and the various clinical manifestations of toxic aquatic origin are often unknown to the public, while even the doctor is barely more conversant with them. To compound the problem, the boom of plane travel and extreme tourist mobility, to holiday clubs and even submarine safaris, has meant that afflictions of an aquatic nature may onset after the tourist's return from far-off waters ("imported" dermatosis). For this reason, an Italian dermatologist may be asked to recognize and treat an unfamiliar disease contracted in the Caribbean or in Polynesia, for instance.

Although many aquatic dermatological diseases resolve spontaneously, they must not be underestimated. Immediate, appropriate treatment of these clinical forms can prevent very serious systemic consequences. Clearly, therefore, clinical suspicion must be followed by a firm diagnosis, especially in the presence of life-threatening reactions.

In this volume, our modest clinical knowledge has been backed up by the data, still by no means exhaustive, reported in the world literature focusing on marine dermatology and aquatic dermatology in general. In addition, many textbooks on aquatic biology have provided a rich source of reference. The aim of this work is to furnish the doctor with a knowledge of the innumerable aetiological factors underlying aquatic dermatitis complaints, together with the various clinical pictures observed and some notions of specific treatments.

Many of the diseases dealt with are skin afflictions caused by Mediterranean flora and fauna, of course, but some aetiological agents from more remote and exotic seas are also referred, with their respective clinical pictures.

First of all, it is important to bear in mind that unlike in other geographical areas, in the Mediterranean and along the Italian coasts in particular, there are no marine species present that are particularly harmful to man. This characteristic sets the Mediterranean apart from all the other seas and oceans on earth. It features a relatively high quantity of dissolved mineral salts (37.7 g/litre). Also, it is really, brilliantly blue, again unlike the oceans and the North Sea, and the water is relatively transparent thanks to the hydrodynamism of the strong sea currents and marked thermal variations. However, the lesser frequency and severity of disease caused by Mediterranean fauna is perhaps the very reason (although no justification) why the biotoxins present in this sea have been so little studied, in comparison with their tropical counterparts, and why so little research has been devoted to the specific antisera.

Diseases caused by marine organisms can be of three different types: 1. toxic, 2. toxo-traumatic, 3. traumatic. This volume concentrates particularly on the first two types, and there are only passing references to traumatic lesions (such as an encounter with a shark, for instance), as these are not properly of dermatological interest. Apart from the diseases caused by biotoxins, the wide spectrum of reactions has been completed by descriptions of those caused by microscopic organisms, that are present not only in the sea but also in swimming pools and aquariums. Finally, a description of exposure to non biotic skin diseases induced by various mechanisms such as contact with salt and freshwater concludes this clinical overview of aquatic dermatology.

2 The aquatic environment and its biotoxins

Fig. 2.1. *Serpula vermicularis* (Polychaeta) and *Phorbas tenacior* (blue sponge)

T he aquatic world, together with its animal kingdom, is renowned for its enchanting beauty. In particular, the vast tropical coral reefs offer a shimmering, glittering underwater panorama featuring an infinite variety of hues, sometimes clashing but always ultimately harmonising. In some cases, however, these beautiful shapes, elegant movements and profusion of colours seem to go hand in hand with disease and death. This is one of the great paradoxes of the aquatic world.

An analysis of the map of the hydrosphere shows that the centre of the world of potentially harmful aquatic animals is the great area of the Indian and Pacific Oceans, although aquatic animals of less aggressive type are present in all the seas. The defence and/or offence mechanisms of marine fauna can be of two different types, physical or chemical (Table 2.1) [1, 2]. Some observations of the latter type have unveiled a fascinating sector of marine biology that is still largely a mystery.

Aquatic biotoxicology is the science that studies aquatic biological toxins, and as this book focuses above all on the skin damage that can be wrought by poisonous or venomous aquatic creatures, it seems wise to start off with an overall classification of toxic aquatic animals [1, 2].

Toxic aquatic animals

The brief classification of toxic aquatic animals listed below includes a large number of species living in enormously different geographical areas and habitats [2].

Invertebrates

Invertebrate toxic aquatic animals, i.e. those with no backbone, can be subdivided into the following groups or phyla.

Table 2.1. Aquatic animals and types of lesions they provoke

1. Aquatic animals inducing mechanical injuries

Sharks
Giant *Manta* rays
Barracuda
Moray eels
Giant groupers
Sawfish
Piranhas
Crocodiles
Alligators
Gavials of the Ganges
Seals
Sea lions
Polar bears
Walruses
Killer whales
Giant squid

2. Venomous aquatic animals: Invertebrates

Porifera (sponges)
Coelenterates (hydroids, sea anemones, jellyfish, corals)
Annellidae (Polychaeta)
Molluscs (cone shells, cephalopods)
Arthropods (aquatic bugs)
Echinoderms (starfish, sea urchins)

3. Venomous aquatic animals: Vertebrates

Venomous fish
 Stingrays
 Catfish
 Moray eels
 Weeverfish
 Scorpionfish
 Surgeonfish
 Flying gurnards
Venomous snakes
 Sea snakes
 Freshwater snakes
Venomous freshwater mammals
 Platypus

4. Poisonous aquatic animals: Invertebrates

Protozoa
Coelenterates (sea anemones, corals)
Echinoderms (sea urchins, sea cucumbers)
Arthropods (crabs, lobsters)

5. Poisonous aquatic animals: Vertebrates

Ichthyosarcotoxic fish
 Ciguatoxic fish
  Surgeonfish
  Triggerfish
  Butterflyfish
  Dolphins
  Wrasses
  Mullet
  Moray eels
  Parrotfish
  Mackerel
  Porgies
  Barracuda
 Clupeotoxic fish
  Herrings
  Anchovies
 Scombrotoxic fish
 Hallucinogenic fish
  Sea chub
 Tetrodotoxic fish
  Puffer fish
  Porcupinefish
  Sunfish
 Ichthyotoxic fish
  Sturgeon
  Gars
  Whitefish
  Catfish
  Codfish
 Ichthyohaemotoxic fish
  Freshwater eels
  Crabs

Ichthyocrinotoxic fish
 Lampreys
 Hagfish
 Pufferfish

Other poisonous vertebrates
 Amphibians
  Salamanders
  Newts
  Frogs
  Toads
 Reptiles
  Turtles
 Marine mammals
  Dolphins
  Porpoises
  Whales
  Polar bears
  Sea lions
  Walruses
  Seals

6. Aquatic animals with an electrical apparatus

Torpedinidae

1. Protozoa
This group includes single-cell planktonic organisms that man can come into contact with when eating molluscs or fish that feed on Dinoflagellates.

2. Porifera
Some sponges (Fig. 2.1) produce chemical substances that are highly irritant to the skin.

3. Cnidaria (*Coelenterata*)
Few species of Coelenterates are poisonous to eat but most of them are venomous.

4. Platyhelminthes
Various species of Platyhelminthes are considered to be poisonous to eat.

5. Annelida
Some Polychaetae worms (Fig. 2.1) are equipped with irritant hairs or spines, while others have venomous glands.

6. Mollusca
Many bivalve molluscs are transvectors of various toxic substances. The cone shell and the octopus are included in this class of molluscs with a venomous apparatus.

7. Arthropoda
Some species of Asiatic crabs are poisonous to eat. There are also a few species of poisonous aquatic insects, belonging to five different families of bugs that inhabit freshwater.

8. Echinodermata
Some species of sea urchins are poisonous, having toxic eggs, and so are some sea cucumbers.

Vertebrates

1. Poisonous and venomous fishes
Many fishes can cause human bio-intoxication when eaten, due to the presence of toxic substances. This class does not include fish that have been accidentally contaminated by pathogenic bacteria. The largest category is that of ichthyosarcotoxic fish, that contain poisonous substances in their muscles, viscera or skin, which obviously cannot be destroyed by heat or gastric juices.

The second major category of poisonous fish is that of ichthyocrinotoxic fish that release toxins through the skin by means of specialised secretory organs. The third category includes the various venomous fish with specialised secretory organs and a wound-producing apparatus, such as spines or teeth.

2. Venomous amphibia
Some amphibians (salamanders, toads, newts) produce very strong poisons.

3. Poisonous reptiles
Some species of sea turtles are considered to be poisonous through eating toxic plants, but the precise source of the poison is unknown. Water snakes make up the most numerous reptile category, and some of these contain very potent poisons.

4. Poisonous mammals
The liver of some whales, polar bears, walruses, seals and sea lions can be toxic.

The functions of biotoxins

Biotoxins, i.e. "poisonous" organic products of bacterial, vegetable or animal origin, are substances that have various different biological actions, of variable severity, when introduced into other organisms. A substance, animal or plant is poisonous if it is harmful to eat (e.g. some mushrooms and some fish are poisonous). Instead an animal is venomous if it produces substances that are harmful when they enter the bloodstream (e.g. the viper and some Coelenterates are venomous, but their secretions are innocuous when eaten). Obviously, the toxic products of a poisonous animal can be harmful when injected [3].

As shown above, many species of marine Vertebrates and Invertebrates produce biotoxins. An animal will be described as "poisonous" or "venomous" according to the use it makes of its poison: when this is used as a defence or offence mechanism and to capture prey for food, the animal is said to be "venomous", or actively toxic. Instead, a "poisonous" or passively toxic animal produces or ingests substances that are harmful when the animal is eaten [3].

There are a large number of data in the literature on the biotoxins of marine animals [4-14]. Some general notions are reported below, while the toxins inherent to each animal species will be dealt with in the relative chapters.

Most marine animal poisons serve to capture prey for food. However, not all marine animals use toxins just to procure food: microphagic animals feed on live or dead organisms and organic waste floating in the water or mingled with the sand, and therefore filter the water or ingest the sand to obtain the nourishing substances. For this reason, some of these animals may be toxin carriers because they act as filters, and any toxic substances present in the micro-organisms or in the aquatic environment will accumulate in their organs and make them poisonous to eat.

Instead, macrophagic animals need to use defence and offence mechanisms to capture and immobilize their prey. Hence, while the mammalian salivary glands have only a digestive function, in many Invertebrates and some Vertebrates (snakes) these glands secrete substances that are actively toxic to the nervous system or other organs of the prey. Cephalopod Molluscs, for instance, immobilize their prey with secretions from their posterior salivary glands, which contain both digestive proteolytic enzymes and biotoxins. Snakes also produce many poisons that are injected with saliva into their victims. Coelenterates capture their prey, such as fish, with their tentacles and immobilize them by injecting toxins through the nematocysts. Not all the action mechanisms of these immobilizing neurotoxins are known: some act on the brain centres or ganglial chains, and others on the peripheral nervous system by impeding conduction or transmission of nerve impulses at the level of the neuromuscular sheath.

Biotoxins may be used purely as defence mechanisms. The scorpionfish (*Scorpaena*) defends itself from predators by means of its venomous dorsal spines. The ray (*Dasyatis*) has a strong, well-developed sting apparatus in the tail which is thrust into the body of the predator. Not only does this organ provoke a painful, lacerating wound, but it also conveys the secretions of the potent poison glands situated at its base.

In short, when they are produced by the salivary glands, biotoxins serve above all to capture prey for food. The biotoxins in the nematocysts of Coelenterates have the same function. Instead, when they are secreted by glands at the base of the spine or radioles, they have a defensive function against other animals. The role of the toxins present in the muscles or ovaries of many marine animals is difficult to ascertain, but it is certain that thanks to these toxins, such animals are poisonous to other animals and to man.

The biochemistry of biotoxins

Up to now, the biochemical make-up of only relatively few biotoxins has been identified, for various reasons: it is difficult to obtain sufficient material for extracting and purifying the poisons, while we have little knowledge of the environmental distribution of many pelagic or deep-sea animals, and no suitable means for capturing them.

The biotoxins whose chemical nature is known are of various types: some are simple amino or phenol derivatives with a low molecular weight or choline esters, or derivatives of steroid or isoquinoline compounds; others are peptides formed by few amino acids or proteins with a high molecular weight.

Generally, a venomous gland produces various compounds with different functions and chemical structures: the nematocysts of the Coelenterates, for example, contain many active substances with high and low molecular weights.

Research on the synthesis and metabolism of biotoxins is still in its infancy. No antidotes to the various poisons have yet been discovered, even for those that can cause mortal epidemics, like mytilotoxin and tetrodotoxin.

References

1. Williamson JA, Fenner PJ, Burnett JW et al (1996) Venomous and poisonous marine animals. A medical and biologic handbook. University of New South Wales Press, Sydney
2. Halstead BW (1992) Dangerous aquatic animals of the world: a color atlas. The Darwin Press Inc, Princeton
3. Ghiretti F, Cariello L (1984) Gli animali marini velenosi e le loro tossine. Piccin, Padova, 7
4. Banner AM (1967) Marine toxins from the Pacific. I. Advances in the investigations of fish toxins. In: Russel FE, Saunders PS (eds) Animal toxins. Pergamon Press, Oxford, 157
5. Der Marderosian A (1968) Current status of drug compounds from marine sources. In: Freudenthal HD (ed) Drugs from the sea. Marine Technological Society, Washington, 19
6. Baslow MH (1969) Marine pharmacology. The Williams and Wilkins Co, Baltimore
7. Bucherl W, Buckley EE (1971) Venomous animals and their venoms. Academic Press, New York
8. Humm HJ, Lane CE (1974) Bioactive compounds from the sea. M Dekker Inc, New York
9. Russell FE, Brodie AF (1974) Toxicology: venomous and poisonous marine animals. In: Mariscal RC (ed) Experimental marine biology. Academic Press, New York, 269
10. Ruggieri GD (1976) Drugs from the sea. Science 194: 491
11. Scheuer PJ (1978) Marine natural products. Academic Press, New York
12. Hashimoto Y (1979) Marine toxins and other bioactive marine metabolites. Japan Scientific Society Press, Tokio
13. Eaker D, Wadström T (1980) Natural toxins. Pergamon Press, Oxford
14. Habermehl GG (1981) Venomous animals and their toxins. Springer Berlin Heidelberg New York

3 **Dermatitis caused by Coelenterates**

Fig. 3.1. *Cotylorhiza tubercolata*

Coelenterata, or "Cnidaria" (from *knidi*, meaning a nettle), are animals with a simple symmetrical radial structure, a mouth that opens out of a single cavity (coelenteron) and a body membrane consisting of two layers of cells (ectoderm and endoderm) separated by an amorphous jelly-like substance (mesoglea). Owing to the symptoms they induce, Coelenterates are also known as "sea nettles" (Fig. 3.1).

These organisms are often very beautiful and have remarkably elegant active or passive movements. They appear vulnerable and inoffensive but in actual fact they are equipped with recondite microscopic weapons running all over the body surface, which are used for defensive and offensive purposes.

The Coelenterate phylum is very prevalent in tropical and subtropical seas. Of the 9,000 known species, about 100 are harmful to man. The species that are toxic for humans belong to the Scyphozoa (*skyphos* = a cup), Anthozoa (*anthos* = a flower) and Hydrozoa classes (Tables 3.1, 3.2). All these species have in common the same types of offensive organule, the production of toxic substances and a mechanism for injecting these substances into their prey.

Coelenterate nematocysts

Coelenterates have thousands of microscopic organisms, called cnidocytes or cnidoblasts (their other name, Cnidaria, derives from this characteristic) running all over the surface of the body and tentacles [1-14]. These dead organules (also described as "nettle cells" or "stinging capsules") contain within the cytoplasm a particular globe-shaped corpuscle. This body, that is used both for defence and offence, is called a nematocyst because it envelops a long, slender filament wound into a spiral.

On contact with a foreign body, a special external receptor (the cnidocilius) is triggered and the cnidocyst violently ejects these nematocysts, which penetrate the body of the victim, opening a gap through which the filament passes and injects its toxins (Fig. 3.2). Strictly ontogenetically, these anatom-

Table 3.1. Phylum of toxic Coelenterates

Class: *Scyphozoa* (jellyfish)	
Species:	*Aurelia aurita*
Pelagia noctiluca	*Chironex fleckeri*
Rhizostoma pulmo	*Chrysaora quinquecirrha*
Cyanea capillata	
Class: *Anthozoa*	
Subclass: Zoantharia	*Calliactis parasitica*
A. Order: Actiniaria (sea anemones)	*Condylactis aurantiaca*
Species:	
Anemonia sulcata	B. Order: Sagartida
Actinia equina	Species:
Adamsia palliata	*Sagartia elegans*
Aiptasia mutabilis	C. Order: Scleractinia (corals)
Class: *Hydrozoa*	
A. Order: Siphonophora	B. Order: Leptomedusa (hydroids)
Species:	
Physalia physalis	C. Order: Milleporina
Physalia utriculus	Species:
Velella velella	*Millepora alcicornis* (fire corals: not true corals)

Table 3.2. Coelenterates known to be toxic. Reproduced with permission from [4]

Schyphozoa		
	Carybdeidae	*Carybdaea alata Reynaud* (Pacific, Atlantic)
		Carybdaea marsupialis (Pacific, Atlantic)
	Catostylidae	*Catostylus mosaicus* (East Australian coasts)
	Chiridropidae	*Chironex fleckeri* (North Australian coasts)
		Chiropsalmus quadrigatus (Pacific, Indian Ocean)
	Cyaneidae	*Cyanea capillata* (North Atlantic, North Pacific, North Sea, Baltic Sea)
		Cyanea lamarcki (North Atlantic, North Sea, Baltic Sea)
	Pelagidae	*Crysaora quinquecirrha* (tropical seas)
Anthozoa		
	Actinidae	*Actinia equina* (Mediterranean)
		Anemonia sulcata (Mediterranean)
		Condylactis aurantiaca (Mediterranean)
	Sagartidae	*Sagartia elegans* (Mediterranean)
	Zoanthidae	*Epizoanthus arenaceus* (Mediterranean)
		Parazoanthus axinella (Mediterranean)
		Palythoa toxica (Pacific)
		Palythoa tuberculosa (Pacific)
		Palythoa mamillata (Pacific)
Hydrozoa		
	Milleporidae	*Millepora alcicornis* (tropical seas)
		Millepora platyphylla (tropical seas)
	Physalidae	*Physalia physalis* (Atlantic and tropical Pacific)
		Physalia utriculus (Pacific)

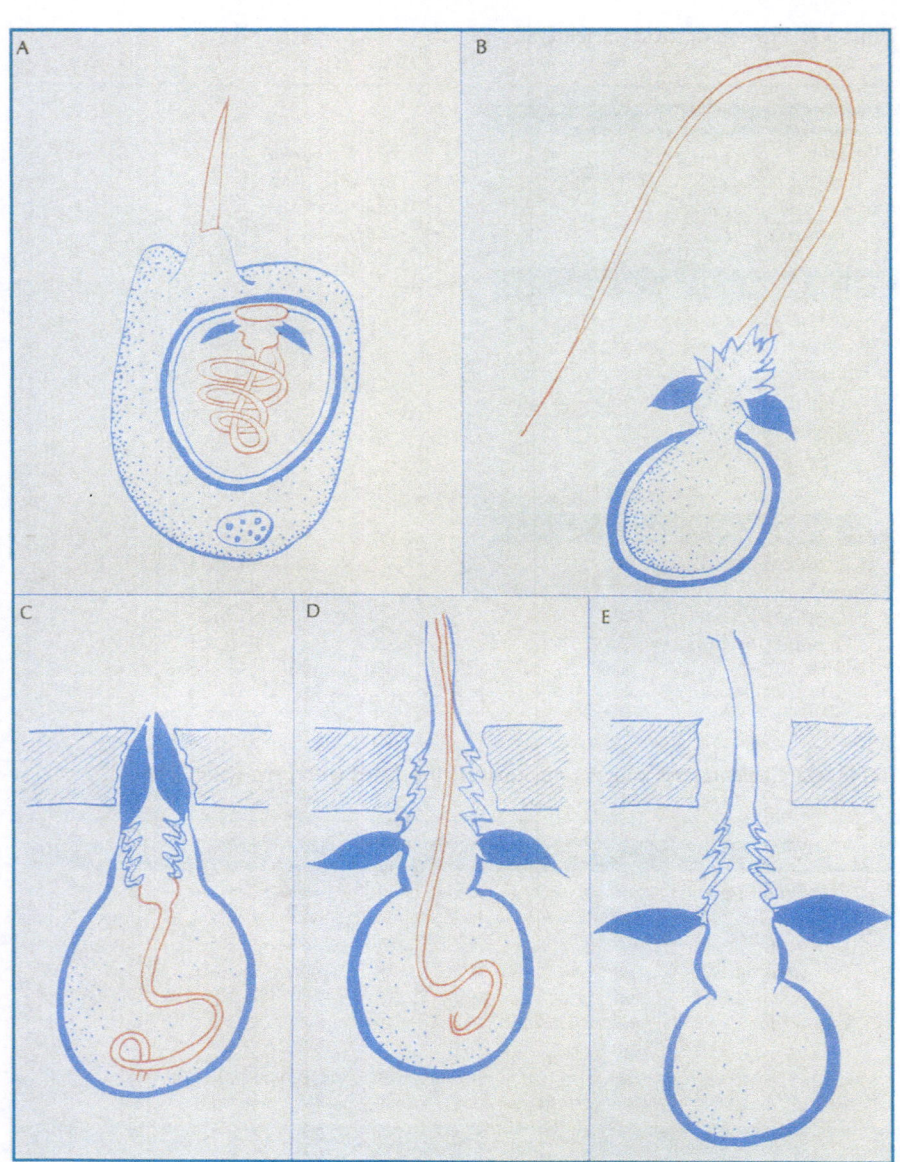

or else floating in the sea carried along by the current or the wind (jellyfish, whose scientific Latin name is Medusa, as their tentacles recall the snakes on the head of the mythical Gorgon).

According to the different species, nematocysts vary greatly in size, shape of the corpuscle and length and morphology of the filament [15]. The latter is not a flagellum but a very slender, flexible hollow tube that is extruded like the finger of a glove when it is thrust outwards. The process of extrusion and penetration of the victim's body has been reconstructed under the electron microscope [16-18], although the underlying mechanisms that trigger expulsion of the nematocysts and extrusion of the filament are not known. Some authors believe that the toxic substances contained in the nematocysts are injected through the filament and others that the toxins are present in the filament itself.

In addition to physical contact with the prey, a specific chemical stimulus stemming from the prey is needed to provoke expulsion of the nematocysts. This stimulus helps to avoid wasting nematocysts, because once the original nematocysts have been ejected others do not reform inside the cnidocytes. The whole discharge process takes less than 3 milliseconds. The thrust force depends partly on the electrical tension generated by the collagen compartment and partly on the intracapsular osmotic pressure, that has been calculated to be about 150 bars. This pressure is determined by the presence of cations (potassium, magnesium, calcium) and some rare polyanions (polyglutamates) in the cnidocyst matrix [19]. In laboratory conditions, expulsion is induced by reduced glutathione, proline, d-methyl-glutathione, valine and leucine, all substances that could probably also be produced by the organisms that have been captured or attacked.

Nematocyst poisons

The nematocysts contain many different biologically active toxic substances, not all of whose chemical structures in the various species are known. In 1902, a French physiologist, Richet, studied the Coelenterate poisons and discovered the anaphylactic phenomenon, which won him the Nobel prize in 1913. Using a glycerinated extract, first of whole tentacles of the *Physalia* and later of the *Actinia*, the author demonstrated its toxic action in some animals, birds and rabbits.

To ascertain the lethal dose, he then injected the extract into dogs, which died after 5-6 days. The animals that had received an insufficient dose of the extract and survived the experiment were used in further experiments. This led to one of the most important discoveries made in the medical field. A dog that had been administered 0.1 ml of glycerinated extract and had not manifested any symptom, was re-injected with a second dose of 0.1 ml of extract after 22 days. A few seconds after this second injection, the animal went into a coma and died after 25 minutes. Richet called this phenomenon "anaphylaxis", or in other words, the reverse of protection [20-22]. Thus, it was discovered that some chemical substances compound, rather than reduce, the organism's sensitivity to their action. From the same tentacles of *Anemonia sulcata*, Richet also isolated three different components: hypnotoxin, thalaxin and congestin. The first induces somnolence followed by respiratory paralysis, the second has an urticarial action on the skin and causes cardiac arrest and the last, which is the one with anaphylactic action, causes vomiting, diarrhoea and gastrointestinal haemorrhage.

In addition to these substances and other proteins, various other substances with a low molecular weight, including tetramethylammonium, adenine, γ-butyrobetaine, histamine and its releasers, imidazyl-acetic acid and 5-hydroxytryptamine, have been isolated in Coelenterates. All these substances and the above proteins are contained in the nematocysts. The substances with a low molecular weight have pharmacological properties but are not harmful in the concentrations present in the organules. Only tetramine, histamine and 5-hydroxytryptamine contribute to the effects on the skin, that consist of burning, erythema and oedema. Instead, the toxic effects are exerted by the protein substances, some of whose amino acid sequences are now known [4].

Some of the cytotoxic and cytolytic effects are caused by damage to the cell membranes, secondary to mitochondrial alterations. In other words, the toxic action exerted by nematocyst poisons has a comparable mechanism to that of calcium-dependent phospholipase [23]. In some species of seas anemones, new protease inhibitors that act against trypsin and chemotrypsin have recently been isolated [24, 25]. Some of the toxins present in common sea anemones in the Caribbean seas seem to be able to block the potassium channels or act as antagonists to the glutamate receptor [26].

Because they contain the above chemical substances, Coelenterates can be considered venomous and actively toxic, while they are not considered poisonous because they are innocuous when eaten. In fact, in some Italian regions (the Veneto), sea anemones are eaten raw as sea food or cooked. They are innocuous because their toxins are inactivated by heat, and in any case they are digested by the intestinal proteolytic enzymes.

Some poisonous Coelenterates live among the coral barriers of the Pacific (Tahiti, Hawaii) and the Caribbean (Jamaica): they belong to the Zoantharia family and the genus *Palythoa* (*P. toxica, P. caribaeorum, P. tuberculosa, P. mamillata*). Palytoxin, the most poisonous biotoxin in the animal world, with the most complex chemical structure ever identified

$(C_{129}H_{223}N_3O_{54})$ [27, 28], was chemically isolated in this genus in 1981. It is a very long chain of carbon atoms with a high content of methyl and hydroxyl groups. These Coelenterates, that are toxic even when ingested by fish and other animals, are very similar to small sea anemones, live in colonies at shallow depths and are shaped like mushrooms about 2-3 cm high. Some fifty tentacles protrude from the apex, forming a crown about 1 cm in diameter. The toxin, a non-protein substance, accumulates in the ovary between March and September during the reproductive period and passes into the eggs. It is not known whether palytoxin is present in the nematocysts as well. The toxin exerts its action on the cardiovascular system and especially on the coronary arteries: at the cellular level it increases permeability to sodium and hence induces depolymerization of the cytoplasmic membrane [4].

Skin reactions

A highly variable range of skin reactions is caused by contact with nematocysts, depending on the size of the area stung and the toxicity of the poison. Subjectively, the symptoms vary from a slight pricking sensation to pain, itching and intense burning. Objective signs are generally of erythemato-oedematous type, featuring more or less bizarre shapes.

Diffuse urticarial reactions are also fairly common, together with anaphylaxis (laryngeal oedema, collapse). In children and extremely sensitive subjects, shock and death can ensue.

Reactions to jellyfish

In the Mediterranean there are 11 species of jellyfish, six of which are harmful to man (Table 3.3) [13, 29]. Among these, *Pelagia noctiluca* (Fig. 3.3) is well known to be toxic to the skin [30], while such an action has

Table 3.3. Jellyfish present in the Mediterranean. Reproduced with permission from [13]

1. Species potentially toxic to man	2. Species not toxic to man
Pelagia noctiluca	*Cotylorhiza tubercolata*
Rhizostoma pulmo	*Discomedusa lobata*
Aurelia aurita	*Drymonema dalmatium*
Chrysaora hysoscella	*Nausithoe punctata*
Carybdaea marsupialis	*Parphyllina intermedia*
Rhopilema nomadica	

Fig. 3.3. *Pelagia noctiluca.* Reproduced with permission from [11]

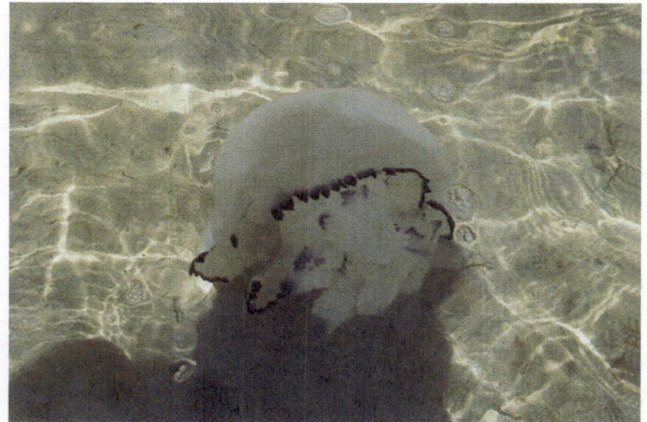

Fig. 3.4. *Rhizostoma pulmo*

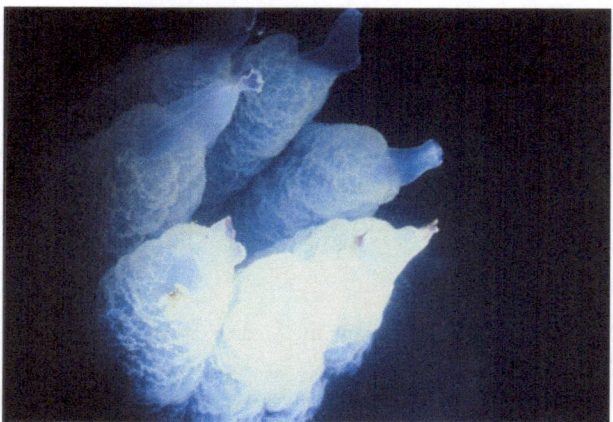

Fig. 3.5. Tentacles of *Rhizostoma pulmo*

still to be confirmed in the following species: *Rhizostoma pulmo* (Figs. 3.4, 3.5), *Aurelia aurita, Chrysaora hysoscella, Carybdaea marsupialis, Rhopilema nomadica.*

Pelagia noctiluca, the "nocturnal jellyfish", a phosphorescent Coelenterate that lives in deep waters, is very common in the Mediterranean, especially in the eastern areas. It is the archetypal jellyfish and is small and the most urticant of the Mediterranean jellyfish. In this species, the cnidocysts are present all over the body surface, not only on the tentacles but also on the bell. In recent years there have been seasonal invasions of *Pelagia noctiluca* along the Adriatic coast, especially in the month of September. Dense colonies float in to a few metres from the shore, causing obvious problems for bathers.

The skin symptoms induced by this Coelenterate are urticarial, painful local lesions that last 1-2 weeks, according to the extent and duration of the contact with the skin and the degree of individual susceptibility. In exceptional cases, more serious pictures can develop, with particularly extensive local reactions and systemic symptoms, up to and including anaphylactic shock [31, 32].

Pelagia noctiluca can cause various clinical pictures induced by both toxic and immunological mechanisms. Recently, Kokelj and Burnett observed three emblematic cases of unusual reactions induced by contact with this jellyfish [30]. The first case was an underwater diver who was stung for the first time by *Pelagia* on the right thigh, which gave rise to a local burning erythemato-vesicular lesion, which regressed within a few weeks. In the following 6 years, the same subject was repeatedly stung on various sites, provoking lesions lasting 10-15 days. Six years after the first episode, a further contact occurred on one side of the face, secondary to a *Pelagia* having entered the mask. This caused not only a local reaction but also the appearance of an intense, burning erythema on the site of the first lesion (the thigh), which resolved within about 10 days.

This particular type of response, that is the result of sensitisation, has also been reported in cases of dermotoxicity induced by insects. In fact, this same unusual reaction has been reported after flea bites, and it has been hypothesized that the poison reacts with the dermal collagen to produce an allergen, towards which the patient's immune reaction is then directed [33]. Allergic dermatitis that has onset on one site, after contact with a sticking plaster (that may contain allergens like colophony, rubber accelerators, lanoline), for instance, can be associated with the appearance of eczematous symptoms in other sites where previous contact with sticking plaster had merely induced an irritant reaction at the time of use.

The second case was that of a child of 11 who was stung on the left arm by *Pelagia noctiluca*. The lesion resolved after about one month, leaving a hypochromic area. In the two years after contact, frequent recurrence of the lesion was observed, in concomitance with any episodes of fever or particular emotional stress. The affliction then gradually regressed spontaneously. Both this patient and the one previously described had significant IgG antibody titres to the crude extract of nematocysts of *Pelagia noctiluca* and also to toxins from *Physalia physalis*, of the Hydrozoa class. The presence of cross-reacting antibodies had already been reported in patients stung by jellyfish [34].

The third case was a doctor who was stung on the left thigh and developed a burning, erythemato-vesicular reaction that healed spontaneously in 15 days. One month after recovery, the lesion recurred with an itching sensation, although there had been no further contact with a jellyfish. This type of recurrent reaction has been reported by other authors [31, 35-37] and we have observed two such cases. Unlike the above two patients, this third patient did not have specific circulating antibodies in serum assayed 19 months after the episode. This means that sensitisation to *Pelagia noctiluca* does not always last long.

Among the other Mediterranean jellyfish, *Rhizostoma pulmo* can sometimes cause dermatitis [38], while it is not known whether *Aurelia aurita* (the "moon jelly"), one of the most beautiful jellyfish with a transparent blue body about 25 cm in diameter, is dermotoxic. It is thought to be innocuous but contact can induce a burning sensation that lasts a few minutes. Instead, the toxicity of *Chrysaora hysoscella* [39, 40], *Carybdaea marsupialis* [41] and *Rhopilema nomadica* [42] has already been demonstrated. This last jellyfish arrived in the Mediterranean in the '80s coming from the Red Sea through the Suez Canal. It is also possible to observe jellyfish that have come in from the oceans to the Mediterranean, especially the west coasts, at some periods of the year.

In comparison with jellyfish coming from the Atlantic, the Pacific and those along the Australian coasts (*Chironex fleckeri* the "sea wasp", *Chiropsalmus quadrigatus*, *Chrysaora quinquecirrha* the "sea nettle", *Carukia barnesi*), the Mediterranean jellyfish are in any case less toxic and only exceptionally induce severe or fatal reactions. *Chironex fleckeri* belongs to the order of Cubomedusae (so-called because of their quadrangular shape) and lives in the Pacific and the Indian Ocean. It is a small, almost transparent Coelenterate that is exceedingly venomous. Among swimmers in Northern Australia, it has been responsible for many cases of death within a few minutes of contact. Three glycopeptide components in its nematocysts have been demonstrated to be responsible for its dermonecrotic, cardiotoxic and haemolytic action, while the lethal effect seems to be due to another neurotoxic component. Another Cubomedusa able to provoke death by the same means is *Chiropsalmus quadrigatus*, present in the seas of the South-East Asiatic. Various toxins with lethal effects have also been isolated from *Chrysaora quinquecirrha* [43-45].

Jellyfish poison consists of a complex blend of polypeptides and enzymes with harmful effects on man due to their toxic and antigenic properties. Human reactions range from local signs and symptoms, to systemic effects and even death. A subject is generally stung when he/she inadver-

Table 3.4. Classification of the reactions induced by jellyfish [7]

1. Local reactions	4. Toxic systemic reactions
Toxic reactions	Malaise
Angioedematous reactions	Asthaenia
Recurrent allergic reactions	Ataxia
Persistent delayed reactions	Cramps and muscular spasm
Distant reactions	Paraesthesia
Contact dermatitis	Vertigo
Urticaria	Slight fever
	Nausea and vomiting
	Irukandji syndrome
2. Local sequelae	
Cheloids	**5. Fatal reactions**
Hyperchromia	Toxic reactions
Hypochromia	Immediate cardiac arrest
Scars	Rapid respiratory failure
Atrophic subcutaneous fat	Delayed renal failure
Gangrene	Anaphylaxis
Contracture	
Eye synechiae	**6. Ingestion reactions**
Glaucoma	Abdominal pain and cramps
Mydriasis	Urticaria
Mononeuritis	
	7. Indirect reactions
3. Successive dermatosis	From aquatic antigens
Herpes simplex	From isolated nematocysts
Anular granuloma	From nudibranchs

tently touches a jellyfish, although contact can occur while handling fishing nets, or touching an animal that has been beached by the current, or even when struck by fragments of tentacles while surfing. The nematocysts expel their filaments with an approximate force of 40,000 g, striking the skin with an estimated force of 2-5 pounds (0.9-2.3 kg) per square inch (2.54 cm) [18, 46]. This energy enables the filament to penetrate the superficial derma, after which the slimy venom adhering to the filament spreads through the circulation and induces the various syndromes (Table 3.4) [35, 36, 46-49].

Jellyfish are widespread in all the seas and are the most common cause of diseases induced by marine animals. The incidence of reactions to jellyfish is not known but is generally presumed to be enormous. Burnett estimates that in Chesapeake Bay (Maryland, USA) there are 500,000 cases of reactions to jellyfish per year, while in Florida they total 60,000-200,000 [7].

The pathogenic mechanism underlying these reactions can be of either toxic or allergic nature, and in the latter case, of either serological or cellular type. Toxic reactions are observed in all subjects after any exposure, and are dose-dependent. Instead, allergic reactions do not onset in all subjects, presuppose a previous exposure to the same toxin, are not dose-dependent, do not necessarily cause exacerbated symptoms at each exposure and can follow contact with toxins from different species of jellyfish belonging to the same class, by means of a cross-reaction.

The toxins conveyed by penetration of the filament into the skin include substances with an enzymatic action (hyaluronidase, collagenase, protease, nuclease, phosphatase) and compounds of quaternary ammonium, catecholamines, proteins, 5-hydroxytryptamine, histamine and its releasers and serotonin and quinine-like products. These are the substances that cause erythema and oedema in acute reactions and contribute to the onset of pain and itching.

Local reactions

Toxic reactions. Contact of the skin or mucosa with a jellyfish induces immediate local pain, that may be due to the reaction of exogenous and endogenous chemical mediators on the sensory nerves of the skin [50]. A special quinine-like mediator present in the poison is likely to be responsible for inducing pain. The pain, that persists for between 30 minutes and 24 hours, is immediately followed by linear skin eruptions of various shapes. These are urticarial lesions that first appear pale, and then rapidly become erythematous (Figs. 3.6-3.8). The lesions last a variable length of time, min-

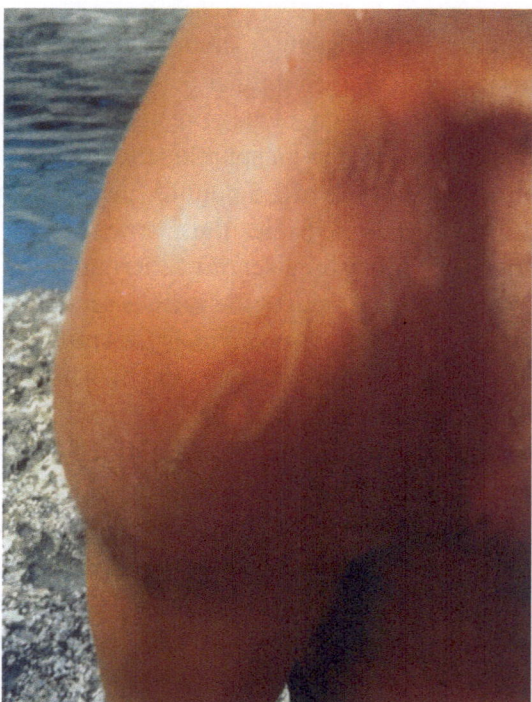

Fig. 3.6. Erythemato-oedematous reaction to jellyfish. The bell and tentacles are reproduced on the skin. Reproduced with permission from [11]

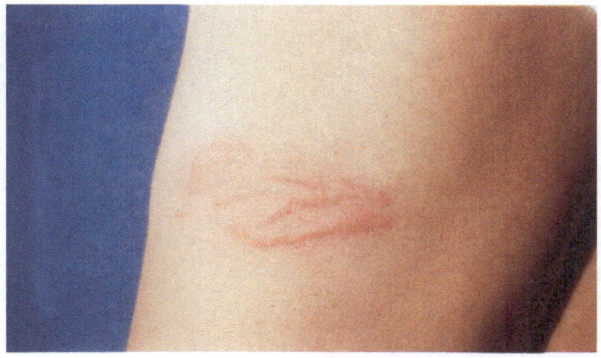

Fig. 3.7. Erythemato-oedematous reaction to jellyfish

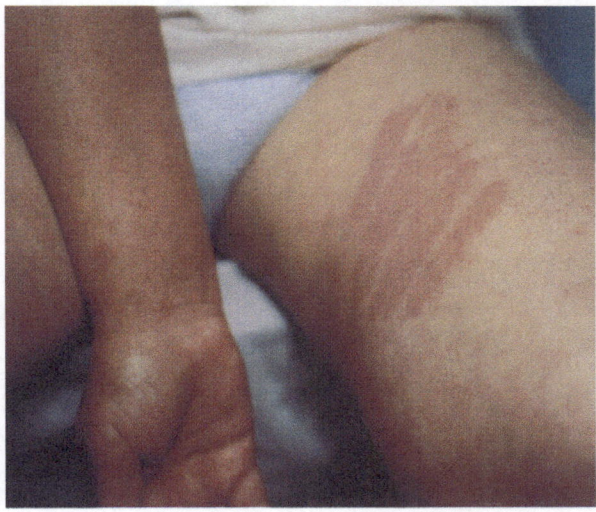

Fig. 3.8. Erythemato-oedematous reaction to jellyfish

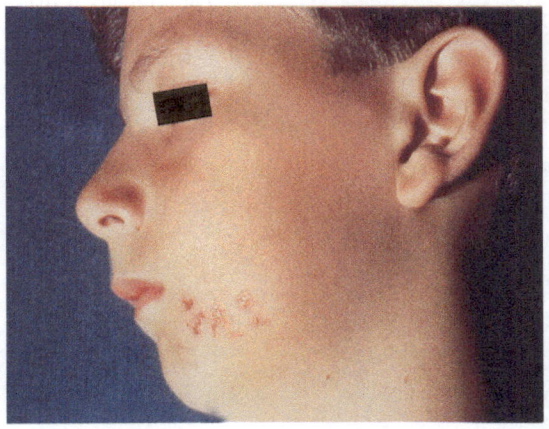

Fig. 3.9. Erythemato-vesicular reaction to jellyfish

utes or hours, but sometimes persist for a longer time according to the intensity of the skin damage.

Lesions can also be vesicular (Figs. 3.9-3.14), blistering (Fig. 3.15), intense-

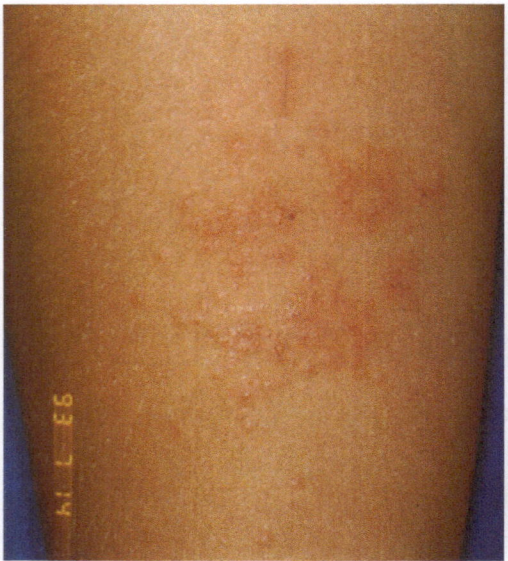
Fig. 3.10. Erythemato-vesicular reaction to jellyfish

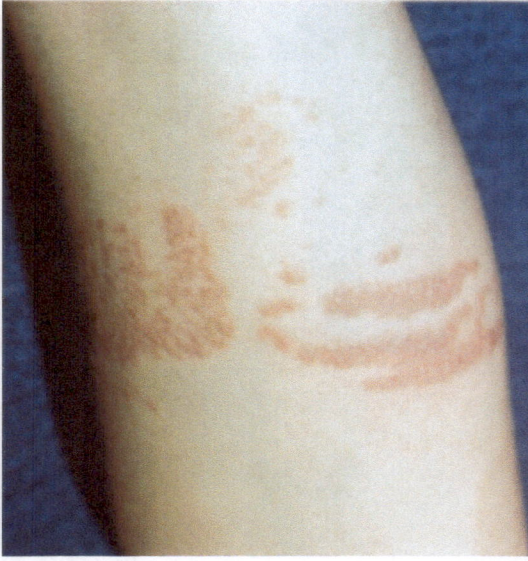
Fig. 3.11. Erythemato-vesicular reaction to jellyfish

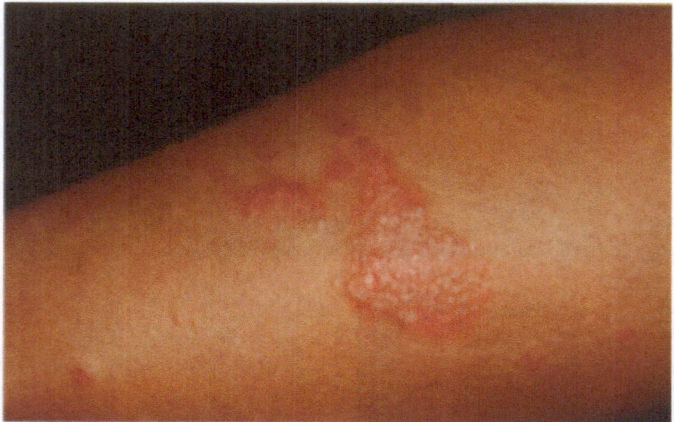

Fig. 3.12. Erythemato-vesicular reaction to jellyfish. Reproduced with permission from [11]

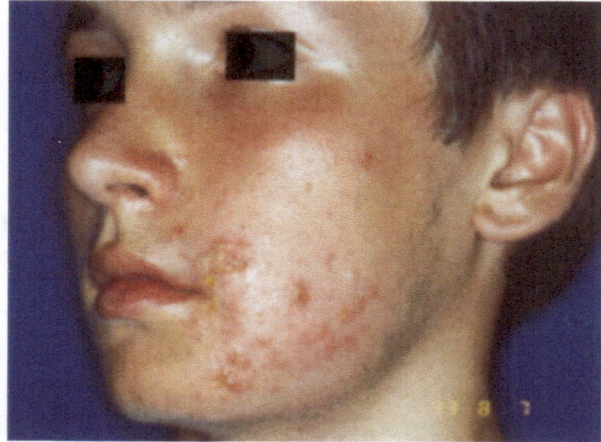
Fig. 3.13. Erythemato-vesico-pustular reaction to jellyfish

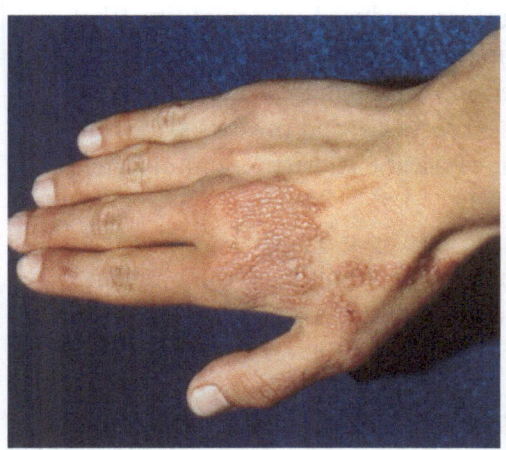

Fig. 3.14. Figured erythemato-vesicular reaction to jellyfish. Reproduced with permission from [11]

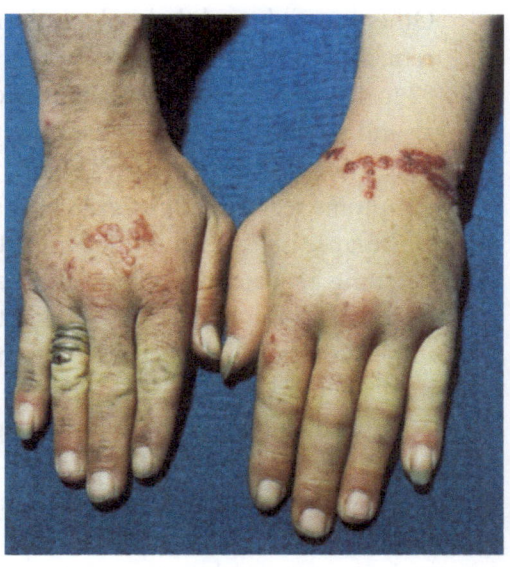

Fig. 3.16. Erythemato-oedemato-vesico-haemorrhagic reaction to jellyfish. The fisherman had plunged his hands into the catch that also contained several jellyfish. Reproduced with permission from [11]

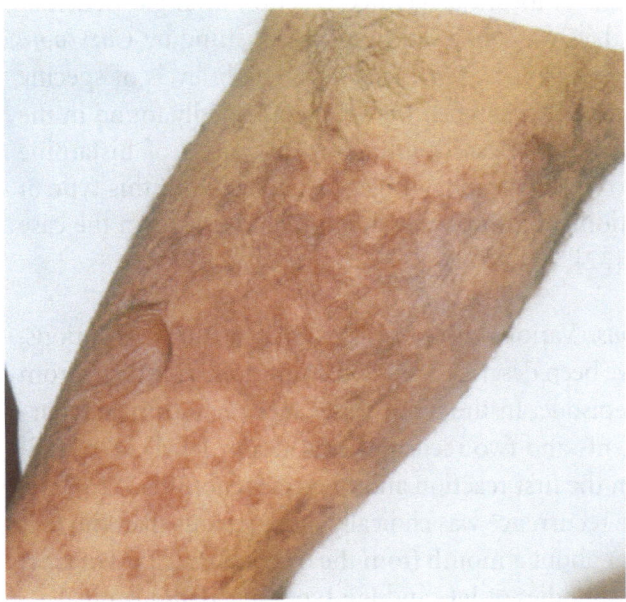

Fig. 3.15. Erythemato-vesico-bullous reaction to jellyfish. Reproduced with permission from [11]

ly oedematous, haemorrhagic (Fig. 3.16) and necrotizing and may be associated with local asymmetrical excessive sweating and successive satellite lymphadenopathy.

Cases of contact with the eyes can induce photophobia, intense pain and burning, conjunctivitis, chemosis, corneal ulceration and palpebral oedema.

Other possible ophthalmological signs are reduced visual acuity (that returns to normal in all cases), iritis, increased intra-ocular pressure (from 32 to 48 mmHg), mydriasis and reduced accommodation. Anterior synechiae and unilateral glaucoma are long term sequelae, and mydriasis, too, can persist for 5 to 24 months. Usually, however, the subjective and objective symptoms regress within 24-48 hours [51].

Because no cases of immune resistance have been reported, a subject can be stung repeatedly with no difference in the resulting symptoms, and injection of the toxins into different animals induces comparable results, some authors initially believed that the clinical reaction was of a toxic rather than allergic nature [46].

Instead, specific human immunoglobulins against jellyfish poison have recently been demonstrated in some subjects, suggesting the existence of a dual pathogenic mechanism, both toxic and allergic, underlying the reactions. In practice, immediate reactions that onset a few minutes after contact are toxic in nature and are the most common types. Allergic reactions are those that manifest in the form of recurrent delayed eruptions without additional causal contact, and are much less common.

Exaggerated local reactions. A local exaggerated reaction of angio-oedematous type lasting 10-14 days was observed in a subject stung by *Chrysaora quinquecirrha* and *Pelagia noctiluca*. For many years, high levels of specific IgE to species with an affinity to jellyfish were then repeatedly found in the patient's serum; his basophils released abundant quantities of histamine when exposed *in vitro* to the *Chrysaora* antigen. A subject with this type of angio-oedematous reaction could probably develop anaphylaxis in the case of any successive stings [32].

Recurrent allergic reactions. Various cases of recurrent linear skin eruptions, itchy but not painful, have been described, that onset after variable times from the first and only sting episode. In the first 5 cases reported, a single recurrence occurred in 4 patients and two recurrent episodes in 1 [31, 35-37, 46-52]. The interval between the first reaction and the recurrence was 7-13 days. In 3 of the 5 patients, the recurrence was clinically more serious than the first reaction. In one case, after about a month from the original sting by *Physalia,* antigen-specific serum antibodies of IgG and IgE type were demonstrated.

In 1987, Fisher reported a case of immediate linear toxic dermatitis that onset on the patient's back and resolved in five days; eight days later, a diffuse erythematous-oedematous eruption appeared in the same site [49]. In 1988, Kokelj and Burnett reported 3 other unusual cases of recurrent eruptions after contact with *Pelagia* [30]. We have observed 2 similar cases: in both there was a recurrence after 20 and 30 days from resolution of the first and only sting episode. Finally, other observations of recurrent eruptions after stings from jellyfish and other Coelenterates have been referred, in one of which oral prednisone had suppressed the primary eruption [46].

These recurrent eruptions are considered to be of an allergic nature in view of the increase in specific serum antibodies. The studies conducted in this field have led to the following considerations: the onset of allergic reactions after jellyfish stings is possible in man; high levels of specific immunoglobulins can persist for many years; recurrences are possible at time intervals ranging from a few days to several months without additional causal contact; a serum cross-reaction to different species of jellyfish is possible; the immune reaction involves both the B and T lymphocyte subclasses.

The finding of increased specific IgE concentrations might help to identify patients at risk or to diagnose stings from animals that the subject was unable to see and describe. In fact, in theory these IgE levels could make any successive episode much more serious, although IgG-blocking antibodies are probably produced in some cases that would mitigate the severity of recurrent reactions.

Persistent delayed reactions. Some persistent delayed granulomatous reactions have been observed after jellyfish stings [36]. At the moment of contact the patient suffers only local pain but after 4-7 days the delayed eruption starts to appear. This will feature nodular lesions of variable size that may persist for months. Histological analysis has demonstrated a dense dermal cellular infiltrate, morphologically similar to the one observed in delayed hypersensitivity reactions.

In view of the histological findings and chronic nature of the eruption, this type of reaction is considered to be an example of delayed cell-mediated immune response. The serum levels of specific IgG and IgE are within normal ranges. Eruptions featuring multiple nodular lesions persisting for several weeks have also been reported after contact with other Coelenterates [36, 52].

Distant reactions. A subject stung on the ankle and right foot by *Physalia physalis* developed an erythemato-oedematous reaction at the level of the ear, the gingival mucosa and the right cheek a few hours later; the eruption lasted 10 days [53]. Seven years later, no serum anti-*Physalia* IgE or IgG were found in the patient's serum.

Contact dermatitis. Contact dermatitis to the tentacles or nematocysts of Coelenterates is an infrequent but possible observation, that can be confirmed by patch test. Such a reaction has been reported to *Physalia* and other jellyfish [46]. A delayed sensitisation reaction was recently observed after repeated contact with *Olindias sambaquiensis*, a transparent hydromedusa about 10 cm long with 380 tentacles. Its usual habitat is tropical and temperate waters between the 23rd and the 42nd south parallel of latitude [54]. It is particularly common along the coasts of the Mar de la Plata and in the Blanca Bay area to the south of Buenos Aires. Patch testing with crude extract of the nematocysts was positive in a subject who had developed a more serious skin reaction only after repeated exposure with slight symptoms. The same patch test was negative in 10 controls [55, 56].

Urticaria. Fifteen minutes after contact with a powdered jellyfish, considered a delicacy, an atopic Japanese patient developed a stinging sensation on his hands. He then developed widespread urticaria with intense itching. By the next day he had a diffuse urticarial eruption, with raised wheals 1-2 cm in diameter, that lasted 30 days. The species could not be identified because the jellyfish had been bought dried [57]. Nine years after this episode, anti-*Chrysaora* IgE and anti-*Physalia* IgG were present in borderline amounts in the patient's serum [46]. Punctiform itchy erythematooedematous lesions onset in other sites a few days after being stung by *Chironex fleckeri* [58].

Histopathological findings. Histological analysis of local skin lesions shows a lymphocyte infiltrate, prevalently in perivasal sites in the superficial derma. These lymphocytes are mainly T helpers (CD4+) and T suppressors (CD8+). Lesions in recurrent forms show a deeper dermal lymphocyte infiltrate, with no deposits of complement or IgG at the level of the dermo-epidermic junction or around the blood vessels [7].

Local sequelae

The local outcome of jellyfish-provoked dermatitis may be constituted by cheloids, post-inflammatory dyschromia (Fig. 3.17) [59], scarring, subcutaneous atrophy, gangrene and contracture [30, 58, 60, 61]. We have most often observed scarring after dermatitis from *Anemonia sulcata*. Cases with vasospasm, contracture and gangrene have been reported in the Indo-Pacific area [62]. Two cases of multiple temporary mononeuritis of a nerve near (but not within) the contact area have also been reported: one after a sting from a corallimorphous anemone in Papua New Guinea, that lasted about 5 months and the other after contact with an unidentified jellyfish in Penang (Malaysia) that also lasted 5 months [63]. Three other cases have been reported, one in Norfolk (Virginia, USA) after contact with an unidentified jellyfish [64], one in Florida after a coral prick [65] and one in Penang (Malaysia) after a jellyfish sting [66]. In all these

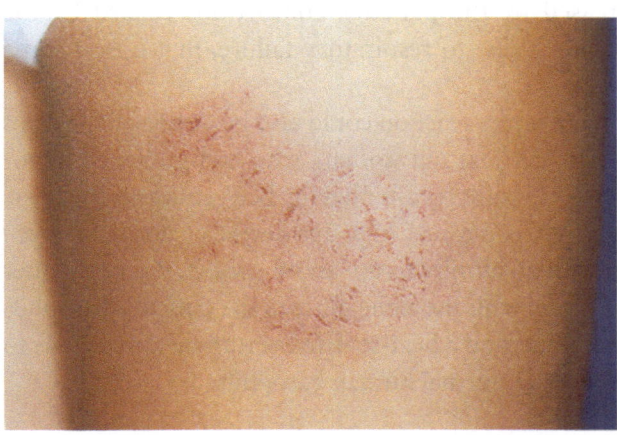

Fig. 3.17. Figured atrophic and dyschromic outcome of reaction to jellyfish

patients, motor and sensory impairment of peripheral nerves and involvement of the nerves near the site of contact were observed but no associated vascular disorders; the complaints resolved spontaneously after several months.

Dermatosis following stings

Dermatitis caused by Coelenterates can be followed by recurrence of herpes simplex or by the onset of an anular granuloma at the site of the sting [46].

Systemic reactions

Skin reactions are sometimes accompanied by toxic systemic symptoms such as malaise, weakness, ataxia, vertigo, cramps and muscular spasm, paraesthesia, nausea and vomiting. A slight fever may also onset.

Severe and fatal reactions

These reactions may be of a toxic or anaphylactic nature. The toxins contained in the poison may exert their lethal effects in different ways. The most potent marine poison known is that of some jellyfish present in the South-East Pacific (*Chironex fleckeri* and *Chiropsalmus quadrigatus*); this induces death in laboratory animals due to cardiac arrest or respiratory failure. In some cases death follows renal failure. It has been calculated that the lethal dose in an adult man would require skin contact with at least 50 foot (15 m) of tentacles of *Chironex*. Perfusion of the damaged skin and spread of the poison are increased by the patient's contortions caused by the intense pain. In man, death is due to respiratory failure, that onsets some minutes after the poisoning.

A fatal or in any case very severe reaction could also be due to anaphylaxis [32]. The serum of a healthy 59 year old woman, who had presented hypotension and absence of pulse 15 minutes after having been stung on the knee, probably by *Pelagia noctiluca*, contained high IgE levels of anti-*Chrysaora quinquecirrha*, a jellyfish with an affinity to the above species, 9 months after the event. Moreover, in contact with the same antigen the patient's basophils released high levels of histamine, and a heat-sensitive serum factor was able to transfer the poison-sensitivity to normal human basophils.

Reactions after ingestion

For the reasons described above, it seems reasonably safe to eat Coelenterates. However, cramps and abdominal pain have been reported, as well as a single case of generalized urticaria brought on by an unknown mechanism [57].

Indirect reactions

There are various possibilities of a jellyfish-induced dermatitis developing without there ever having been direct contact with the animal [49]. Coelenterates can release poisonous antigenic substances into their aquatic environment and these substances can induce sensitisation in swimmers even without any contact with the nematocysts. In any later contact with a jellyfish, the sensitised subject may develop an allergic dermatitis of a serious nature.

When strong storms are blowing, nematocysts detach from the tentacles of jellyfish in particular, float away and continue to release toxins for several months. Contact with these can cause the development of a "dermatitis caused by nematocysts" without any contact with the Coelenterate. In fact after a storm, "epidemics" of moderately itchy skin eruptions can be observed, that are difficult to diagnose in the absence of linear lesions and a history of contact with a jellyfish.

Two species of Nudibranchs (Molluscs) (Fig. 3.18), *Glaucus atlanticus* and *Glaucus glaucilla*, feed on the tentacles and nematocysts of *Physalia;* these nematocysts are not digested but migrate and are stored in their dorsal papillae. Swimmers coming in contact with these "armed" Nudibranchs can be stung by the nematocysts. The ensuing reaction, known as "dermatitis caused by Nudibranchs", is in fact a dermatitis caused by nematocysts.

Fig. 3.18. *Flabellina ischitana* (nudibranch mollusc)

Sex-linked susceptibility

Jellyfish poisoning syndromes are prevalent in the female sex, despite the fact that the number of male swimmers is presumably higher [7]. Most subjects with persistent skin lesions and atrophy of the subcutaneous fat are women. The true significance of this fact may only be clarified by time and accurate epidemiological studies.

Irukandji syndrome

This syndrome is named after an Australian aboriginal tribe in which it has been reported several times [67]. The aetiological agent was then discovered to be the *Carukia barnesi* jellyfish, and the syndrome can now be called carukiasis [68-70]. One hour after contact with the animal, without any skin symptoms being present, systemic symptoms onset, including pain at the site of the sting, often accompanied by intense headache, backache and joint pain, and frequently nausea and vomiting. There may also be fever, cardiac arrhythmia and coughing. The syndrome resolves spontaneously after 1-2 days.

Immunological findings

Jellyfish poisoning may be followed by serological and cellular immunity. IgG and IgM may be produced and persist for many years [71]. In subjects with recurrent forms, the release of circulating lymphokines, leukocytic stimulation or suppression of the natural killer cells have been demonstrated. There can also be antibody cross-reactions to jellyfish of different species. In a healthy volunteer, an intradermal injection of a high dose of toxins gave rise to systemic immune suppression [72, 73].

Reactions to sea anemones

The Actiniaria (*actis + inos* = a ray) belong to the Coelenterate phylum and Anthozoa class (Table 3.1). All species have nematocysts. Owing to their great profusion of colours, these animals often resemble anemone flowers, which is why they are commonly called "sea anemones", "sea roses" or "sea daisies" (Fig. 3.19).

The common species of sea anemones in Italian seas are *Actinia equina, Condylactis aurantiaca, Adamsia palliata, Aiptasia mutabilis, Calliactis parasitica* and above all *Anemonia sulcata*. These various sea anemones cover the sea bed with their tentacles and display iridescent colours ranging from red through orange to purple. *Anemonia sulcata* the "wax rose sea anemone" is common in shallow waters and up to depths of 10 metres; younger examples can frequently be found in pools (Fig. 3.20) and under the tide-line (Fig. 3.21) where they expand to cover the submerged rocks completely. As they grow bigger they creep slowly out towards deeper waters.

Like jellyfish, sea anemones have a transparent body due to a very high water content, over 95% of their body weight. They generally live attached to the sea bed and outside the water they lose their shape and appear as a jelly-like blob. Sea anemones have a highly variable morphology, looking rather like fleshy flowers on a thick stem, with a crown of brightly coloured tentacles issuing from the apex. The tentacles are long and slender and arch downwards like water from a fountain; they wave gently in the water drawing very elegant figures and fantastic arabesques. In fact, sea anemones offer lovers of underwater diving a sight of rare and incomparable beauty (Fig. 3.22).

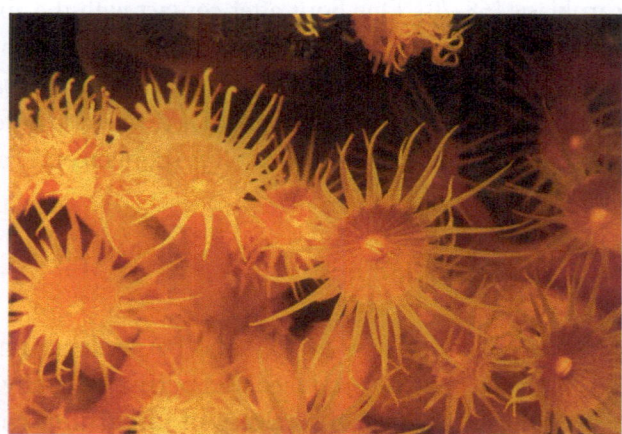

Fig. 3.19. *Parazoanthus axinellae* (sea daisies)

Fig. 3.20. *Anemonia sulcata* in a seawater pool

Fig. 3.21. *Anemonia sulcata*

Fig. 3.22. A field of sea anemones (*Anemonia sulcata*)

From the biopharmacological point of view, too, sea anemones are of particular interest: studies of their tentacles have led to the discovery of anaphylaxis [20-22] and the identification of the active pharmacological substances hypnotoxin, thalaxin and congestin. Thalaxin has an intensely urticarial action on the skin as it can induce the release of histamine in the tissues.

In 1974, equinatoxin was isolated from *Actinia equina*, the "strawberry sea anemone", a very common species in shallow waters in the Mediterranean. Equinatoxin is a polypeptide consisting of 49-53 amino acids with biological and pharmacological properties similar to those of snake venom [74]. In recent years, four toxic polypeptides from *Anemonia sulcata* have been isolated and characterized (the amino acid sequence has also been identified for three of them), all of which have a paralysing action on crustaceans, fish and mammals. Toxic polypeptides consisting of 49 and 51 amino acids with the same action have also been isolated from *Condylactis aurantiaca* [4].

Dermatitis caused by sea anemones

Unlike contact reactions to jellyfish, although those to sea anemones are well known they have rarely been reported in the literature [1, 2, 5, 6, 8, 10-12, 75, 76]. We have had occasion to observe many cases of reactions to *Anemonia*, some of which onset spontaneously after accidental contact with the animal, while others are provoked. The latter pictures are commonly observed in children and young people who, despite being aware of the urticant properties of the sea anemone, play at throwing them at one another (Figs. 3.23, 3.24).

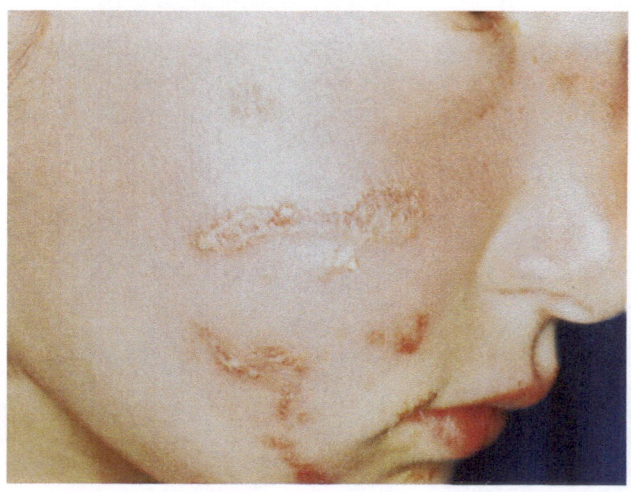

Fig. 3.23. Figured erythemato-vesicular reaction to sea anemone

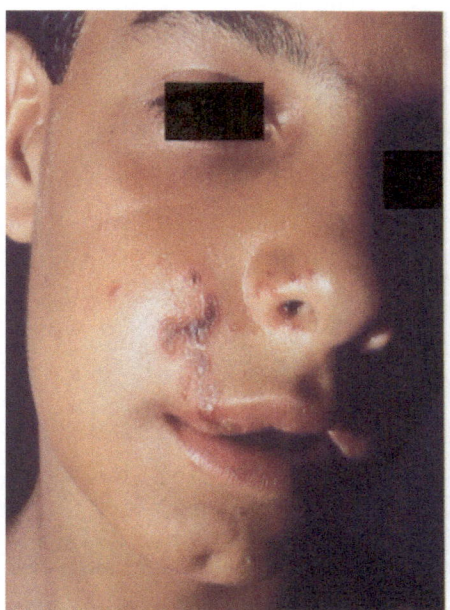

Fig. 3.24. Necrotizing reaction to sea anemone

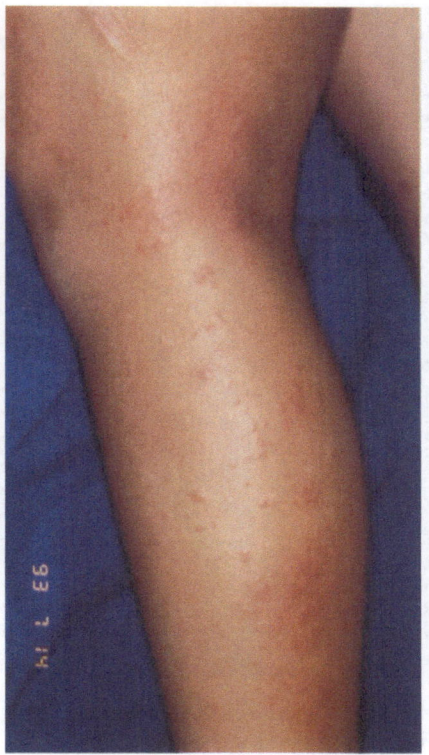

Fig. 3.25. Erythemato-vesicular reaction to sea anemone on a typical site

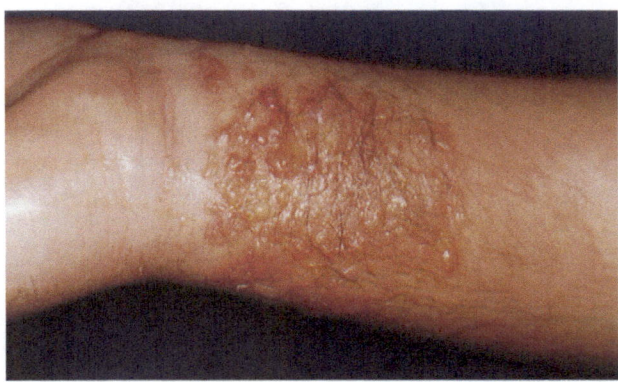

Fig. 3.26. Erythemato-bullous reaction to sea anemone on a typical site. Reproduced with permission from [11]

From the clinical point of view, sea anemones can induce the same clinical pictures as those described above for jellyfish (Table 3.4). Local reactions are most commonly toxic, and in our experience generally present with much more marked symptoms than those due to local reactions to jellyfish. First of all because the lesions are much more extensive, assuming very bizarre pathognomic pictures, notably elegant arabesque-like stripes. Morphologically, in addition to the erythemato-oedematous aspect, the lesions are more often vesicular or blistering (Figs. 3.25-3.34) and some-

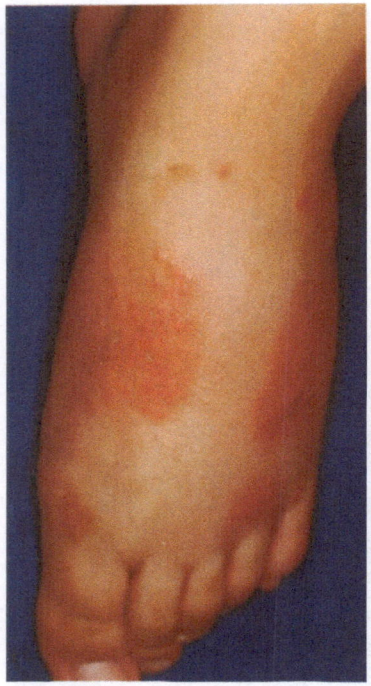

Fig. 3.27. Figured erythemato-vesicular reaction to sea anemone after wading in a pool with open strip sandals

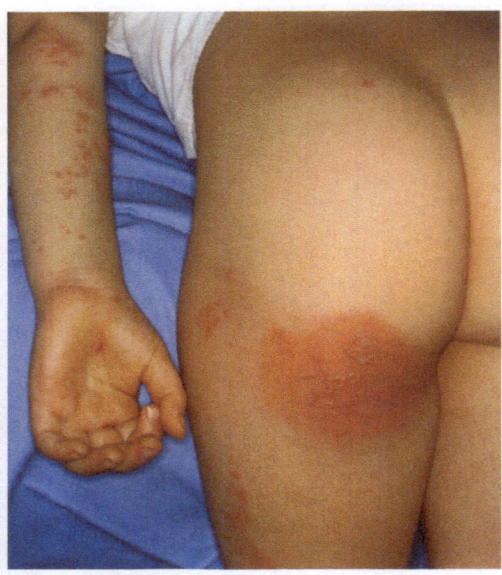

Fig. 3.28. Erythemato-oedemato-vesicular reaction with multiple foci on typical sites. The child had sat on a submerged rock covered in sea anemones

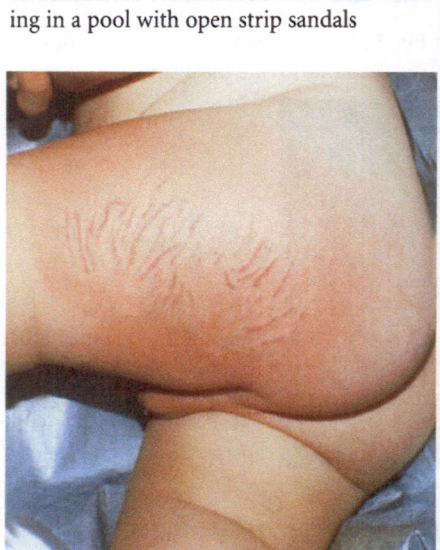

Fig. 3.29. Figured vesicular reaction to sea anemone, with erythemato-oedematous base on a typical site

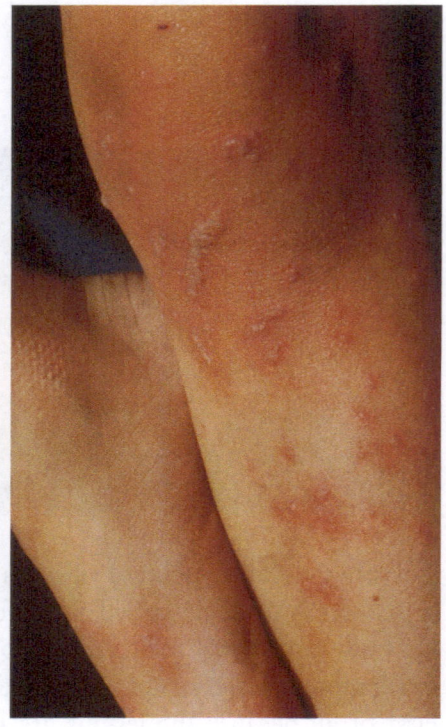

Fig. 3.30. Figured vesiculo-bullous and pustular reaction to sea anemone. The patient had knelt on a submerged rock covered in sea anemones

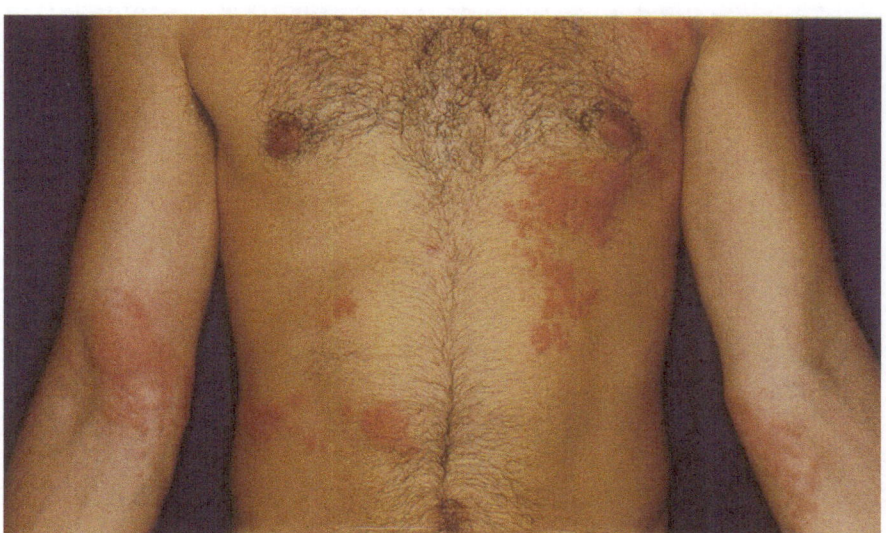

Fig. 3.31. Erythemato-vesicular dermatitis with foci disseminated all over the body in a skin diver who had swum through a field of sea anemones

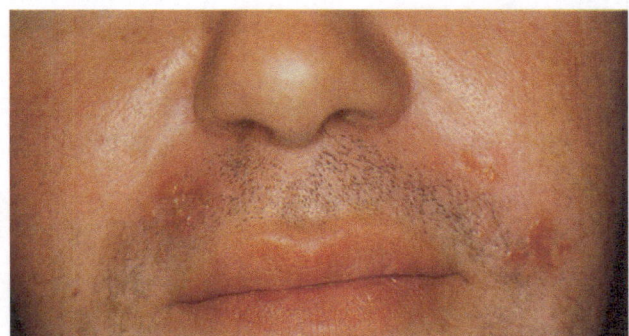

Fig. 3.32. The same case as in Fig. 3.31. Erythemato-vesicular foci of herpes simplex type. Reproduced with permission from [11]

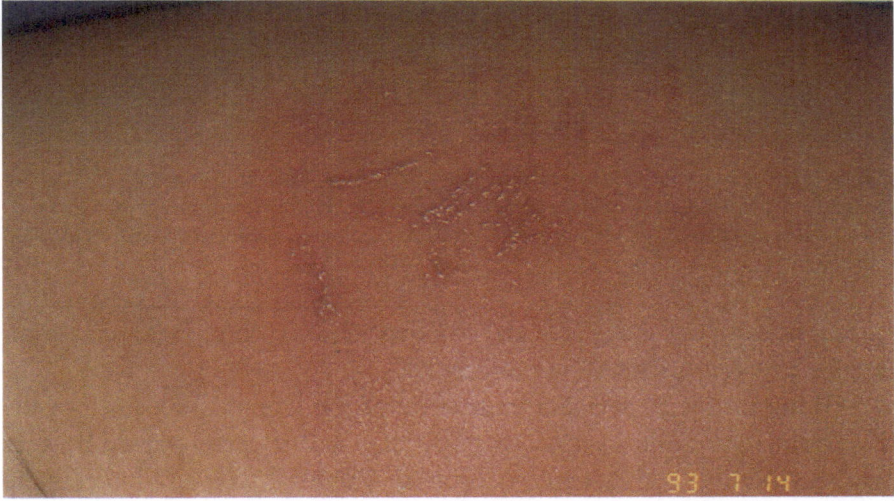

Fig. 3.33. Rounded erythemato-oedematous patch with central figured vesiculation typical of reactions to sea anemones. Reproduced with permission from [14]

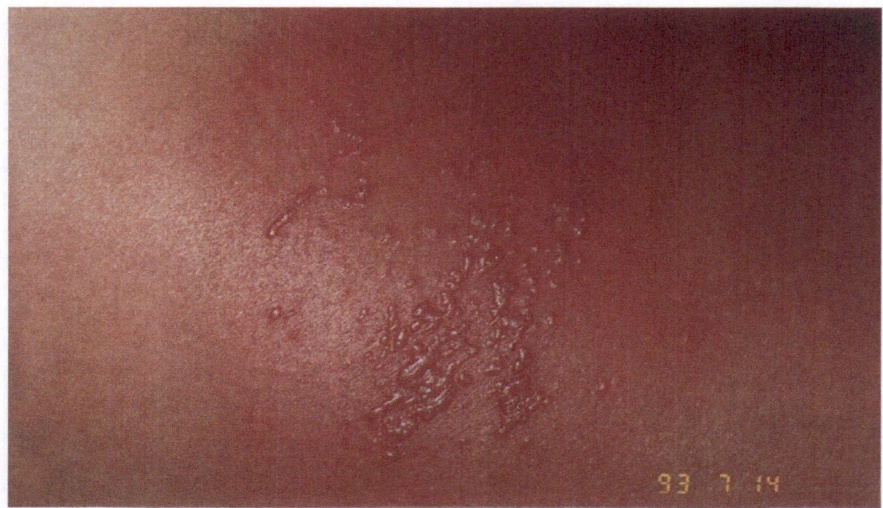

Fig. 3.34. Rounded erythemato-oedematous patch with central figured vesiculation typical of reactions to sea anemones. Reproduced with permission from [11]

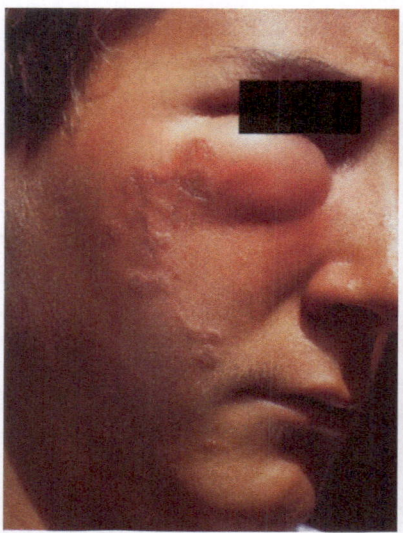

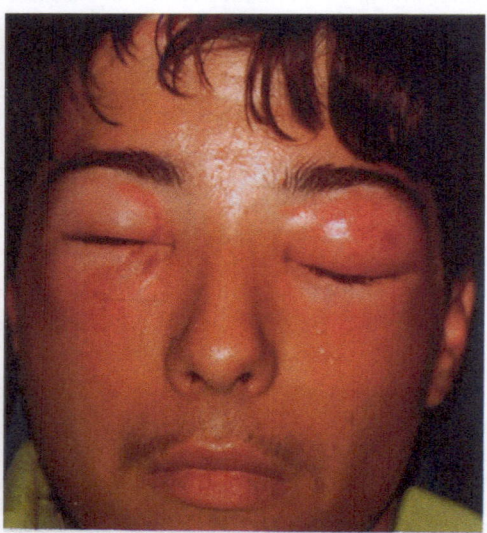

Fig. 3.35. Figured vesicular reaction with severe palpebral oedema from *Anemonia sulcata*. Reproduced with permission from [11]

Fig. 3.36. Angioedematous contact reaction to *Anemonia sulcata* from swimming through a field of these anemones. Reproduced with permission from [11]

times necrotizing. The oedema is often serious enough to create an angioedematous picture (Figs. 3.35, 3.36).

Owing to this greater clinical severity of the lesions, the course of the complaint lasts from 15 days to 20-30 days and is accompanied by very severe subjective and systemic symptoms. The local pain and burning are sometimes intolerable and systemic reactions, such as malaise, weakness and muscular cramps, are nearly always present (Table 3.5). Dyschromic or

Table 3.5. Differential characteristics between reactions to jellyfish and sea anemones

	Reactions to jellyfish	Reactions to sea anemones
Incidence	Frequent	Less frequent
Age	All ages	All ages, especially children
Provoked reactions	Exceptional	Frequent in young people (sea anemones are easily picked up)
Type of contact	Generally superficial	Generally very close
Method of contact	Generally brushing against it while swimming	Generally close when sitting or lying on the rocks
Sites	All sites	All sites, especially posterior face of the thighs, back and volar surface of wrists for above reasons
Extension of dermatitis	Slight (contact with jelly)	Extensive (contact with various areas) for above reasons
Clinical picture	Generally slight	Generally severe
Morphology of lesions	Figured, mainly linear figures	More bizarre and arabesque-like
Clinical lesions	Erythema, oedema, rarely blisters and necrosis	Erythema, severe oedema, frequent blisters and necrosis
Rounded lesions	Not observed	Frequently observed, vesico-bullous at centre and erythemato-oedematous at distinct margins
Local subjective symptoms	Generally pain and slight burning sensations	Generally intolerable pain and burning sensations
Systemic symptoms	Possible, generally slight	Virtually constant and severe
Clinical course	Few days	15-30 days
Sequelae	Infrequent	Frequent
Allergic reactions	Possible	Possible

scarring sequelae are much more common after reactions to sea anemones than to jellyfish (Figs. 3.37-3.40).

No cases of fatal reactions to sea anemones have been reported in the literature. We have observed two cases of systemic reactions [76]. The first was a 9 year old boy, observed in July 1987. He presented an erythemato-oedematous dermatitis with unusually configured lesions on the flexural surface of the right thigh. The lesions were an agglomerate of erythematous, oedematous and vesico-bullous stripes; the erythema, which was bright red with a purpuric imprint, had fairly distinct outlines. The stripes varied in length from 2-7 cm and were interwoven, creating an elegant abstract design with a central core and branches spreading out in various directions, rather like pseudopods (Fig. 3.41). Identical lesions with a more elementary design

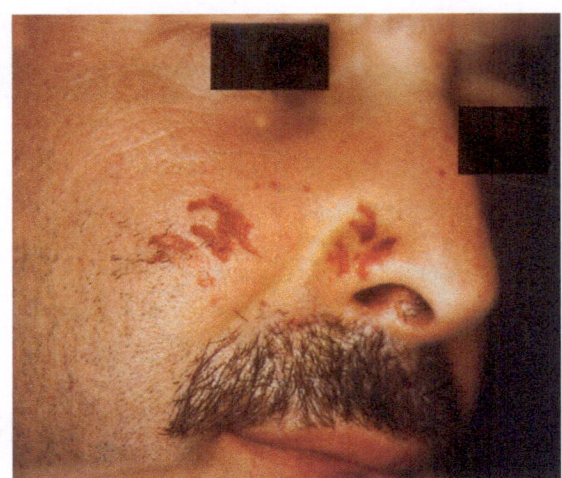

Fig. 3.37. Figured scabs and atrophic outcome of reaction to sea anemone

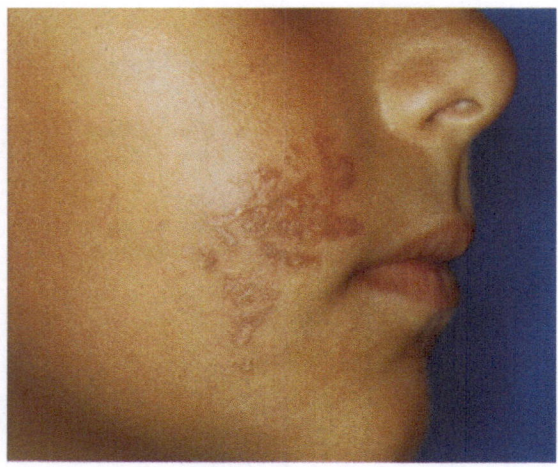

Fig. 3.38. Figured atrophic and dyschromic outcome of reaction to sea anemones. Reproduced with permission from [11]

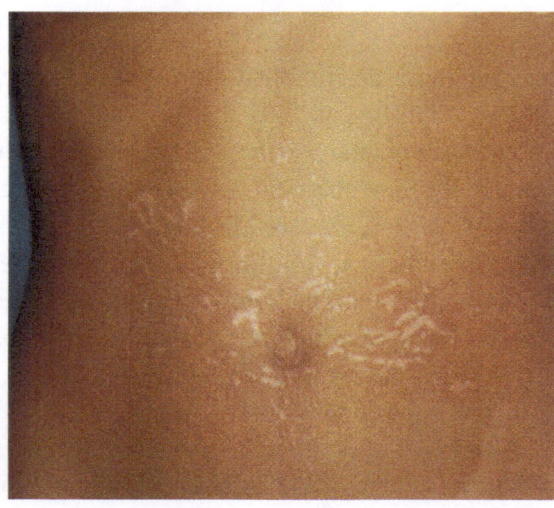

Fig. 3.39. Figured atrophic and hypochromic outcome of reaction to sea anemones. Reproduced with permission from [11]

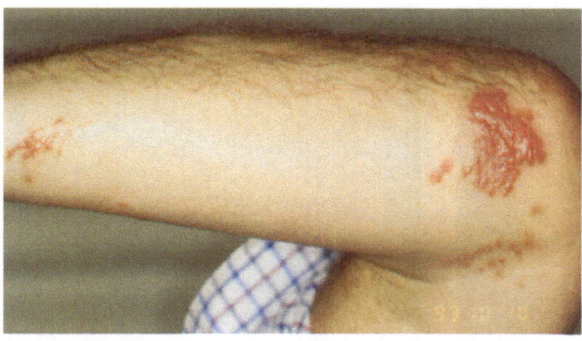

Fig. 3.40. Cheloid outcome of reaction to sea anemone. Reproduced with permission from [14]

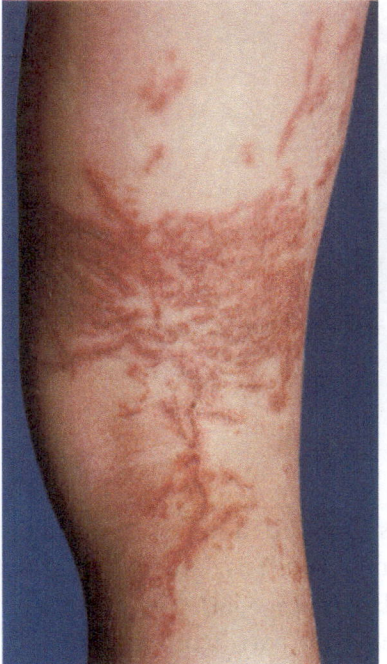

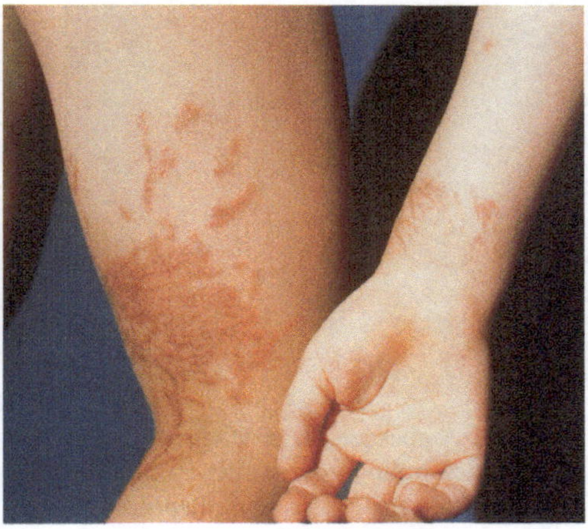

Fig. 3.42. The same case as in Fig. 3.41 with comparable lesions on the volar surface of the wrist. Reproduced with permission from [14]

Fig. 3.41. Arabesque-like erythemato-vesicular dermatitis from *Anemonia sulcata* on typical sites. Reproduced with permission from [11]

were present on the volar surface of the right wrist (Fig. 3.42). The complaint provoked intense pain and burning and was associated with headache, nausea, vomiting, bronchospastic crises and muscular cramps.

The skin complaint and subjective and systemic symptoms had onset at the seaside near Bari, a few minutes after the child had sat down on a partly submerged rock. On examination of the place we collected (with gloved hands) the jelly-like bodies of a few sea anemones that were identified by the Laboratory of Marine Biology of Bari University as members of the *Anemonia sulcata* species.

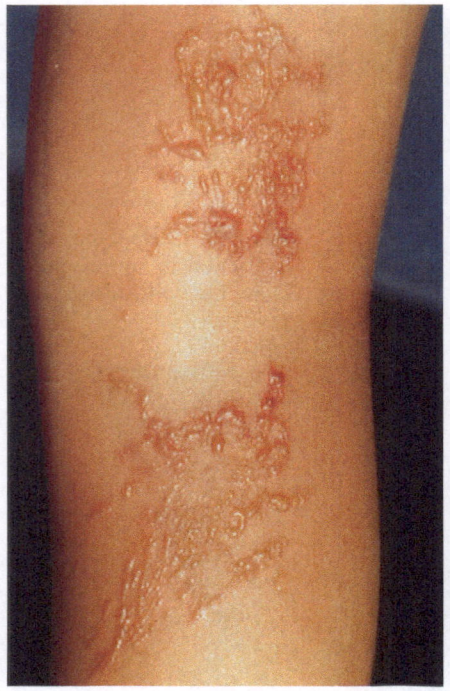

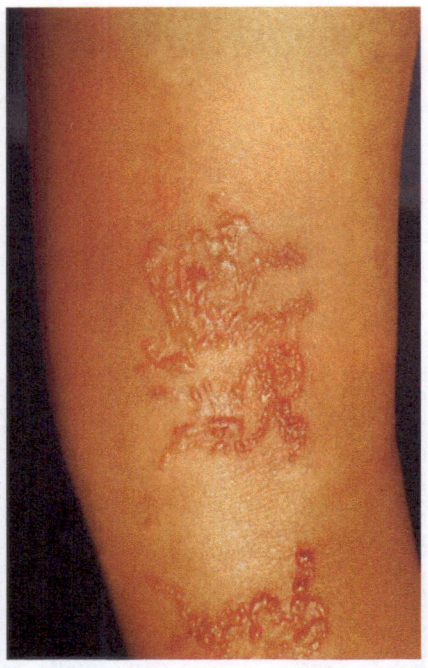

Fig. 3.44. The same case as in Fig. 3.43

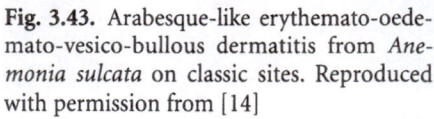

Fig. 3.43. Arabesque-like erythemato-oede-mato-vesico-bullous dermatitis from *Anemonia sulcata* on classic sites. Reproduced with permission from [14]

The second case was a girl aged 12 years observed in July 1988. She presented with an extensive erythemato-oedemato-bullous figured dermatitis on the posterior face of the right thigh (Figs. 3.43, 3.44). The complaint was associated with intense pain at the affected part, nausea, vomiting, headache, muscular cramps and bronchospastic crises. The skin complaint and systemic symptoms were again due to brief contact with the *Anemonia sulcata* species. In both cases observed, the dermatitis resolved in about 25-30 days.

Sagartia's dermatitis

One of the most common reactions to sea anemones is dermatitis from *Sagartia elegans*, also known as "sponge fishermen's disease" or "maladie des pêcheurs d'éponges nus", as it is called in some parts of the Mediterranean. *Sagartidae* are very common Coelenterates that live symbiotically at the base of sponges, and are to be found from Iceland right down to the southern Mediterranean. They are shaped like flowers about 1-4 cm long and have a cylindrical polypoid body with two rows of tentacles arranged radially.

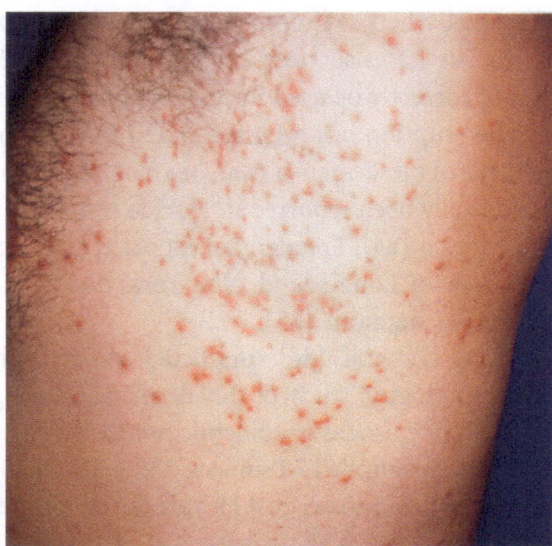

Fig. 3.45. Seabather's eruption: diffuse erythemato-papulo-vesicular dermatitis after swimming in the Caribbean sea. Reproduced with permission from [87]

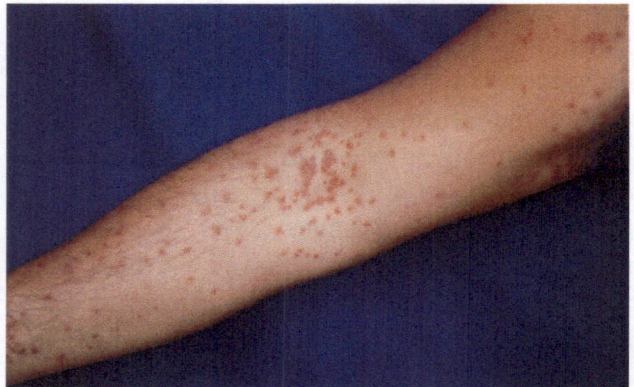

Fig. 3.46. The same case as in Fig. 3.45

Fishermen harvest these sponges with their bare hands, remove stones and other encrustations from the base and put them in a net slung round their necks. During these manoeuvres the fishermen come into contact with *Sagartia*'s tentacles and after a few minutes they feel burning and itching sensations, which are followed by erythema and blisters. The erythema is bright red at first but then turns purple. The dermatitis may be associated with systemic symptoms, including headache, nausea, vomiting, fever, shivering, muscular spasm and collapse. The skin complaint takes quite a time to resolve and sometimes multiple abscesses develop, which may evolve into ulcers [1, 77].

Seabather's eruption

The first description of the sea bather's eruption was of a papulous, itchy complaint (Figs. 3.45, 3.46) on skin areas covered by the costume or wetsuit,

affecting bathers swimming on the south-eastern coasts of Florida [78]. The dermatitis has also been inappropriately labelled "sea lice", a name that should correctly be used for the metazoal parasites of fish [79].

The eruption has been attributed to the nematocysts of Cnidaria (jellyfish, Portuguese man-o'-war, sea anemones, hydroids and corals). It has periodically been reported in Florida [80, 81], Cuba [82, 83], the Caribbean and Mexico [84]. Freudenthal reported cases of seabather's eruption along the coasts of Long Island, New York, due to the larvae of the *Edwardsiella lineata* sea anemone [85].

Wong and coll. [86] conducted an interesting aetiological and clinico-histopathological study of this dermatitis during the spring and summer of 1992, when episodes of epidemic proportions occurred in south-east Florida. The authors enrolled 70 subjects (36 male and 34 female) who were resident or holidaying in south-east Florida during the period between April and June 1992. It was the first experience of dermatitis for 76% of them, while the remaining cases had a positive history. Fairly severe symptoms were present in 22 patients. Most of them had been swimming at the time of the event (78.6%), while the others were underwater diving (8.6%), surfing (7.1%) or boating (7.1%). About 25% of the patients developed the symptoms as they came out of the water, while in the others the clinical signs appeared after a mean of 12 hours. Itching, generally intense, was present in 69 subjects. Other subjective symptoms included malaise and fatigue (23%), fever (18.6%), shivering, headache, nausea, coughing, abdominal pain and diarrhoea.

The objective lesions consisted of many erythematous papules very near together, that sometimes evolved into pustules or blisters. Some of these lesions had a follicular distribution. In 11% of the cases there were also urticarial lesions. The eruption was prevalently monomorphous and lasted about 3 days. Regional adenopathy developed in 10% of patients. The papules were more numerous and inflamed in those who had kept their bathing costumes on after coming out of the water. In 68% of subjects, the lesions were confined to covered sites, while they were present in both covered and uncovered sites in the others. The lesions were concentrated under the swimming costume, in the skin folds and in sites where the costume hugs the skin closely. The eruption lasted from 1 to 4 weeks (a mean of 12.5 days). The histological findings were not diagnostic: the most common picture was of a superficial or deep infiltrate, prevalently perivasal, of lymphocytes, eosinophils and neutrophils. In some cases suppurating folliculitis or subcorneal pustules were evident; there was no spongiosis, necrosis or alteration of the sub-epidermal membrane.

Analysis of the waters showed that the cause of the eruption was larvae of the *Linuche unguiculata* jellyfish ("thimble jellyfish"). The serum of three patients showed high titres of specific IgG to this jellyfish. General and topical treatment was not very effective, although in some cases par-

tially successful results were obtained with corticosteroids administered topically or systemically.

Other authors have also reported afflictions in which *Linuche unguiculata* was the causal agent [81, 88, 89]. It is not known why the episodes in south-east Florida are periodical but it seems likely that they may be caused by the Gulf Stream: the particularly high temperature of the water in these zones in some periods, such as summer 1992, might offer optimal reproductive conditions for this jellyfish and hence an increased quantity of larvae. This makes it likely that seabather's eruption may be caused by different Coelenterates in different waters, giving rise to the same clinical symptoms. From the pathogenic point of view, the reaction may be of a toxic or allergic nature, as shown by the findings of specific human immunoglobulins. Differential diagnosis must be made between seabather's eruption (the Caribbean coast of the central Atlantic, in covered skin areas and due to Cnidaria larvae) and dermatitis caused by Cercariae (that are ubiquitous especially in freshwater and affect exposed skin areas) and by seaweed (in Hawaii, fresh and saltwater seaweed, affecting covered skin areas) (Table 3.6).

Reactions to physaliae

Physaliae (*physaleos* = full of air) are floating bluish-purple jellyfish that usually live in the tropical regions of the Pacific, Atlantic and Indian Ocean. They are wafted along by the tides and surface winds and sometimes end up on the European Atlantic coasts and in the Mediterranean. In recent years, in fact, a large number of physaliae have been seen on the Mediterranean coasts due to the warm temperatures that favour their development, the strong winds from the south-east that push them towards our shores and the absence of opposing winds.

Members of the Cnidaria, physaliae belong to the Hydrozoa class (Table 3.7). The Physalia genus includes two species: *Physalia physalis,* the most representative type (present in the tropical Atlantic and the Mediterranean) and *Physalia utriculus,* present in the Indo-Pacific region and South of Japan.

Table 3.6. Differential diagnosis among seabather's eruption (BE), dermatitis from cercariae (DC) and dermatitis from seaweed (DS)

Factors	BE	DC	DS
Type of water	Salt	Salt and fresh	Salt and fresh
Areas affected	Covered and uncovered	Uncovered	Covered
Cause	Cnidaria larvae	Schistomiae	*Lyngbya majuscola*
Geographical areas	Florida, Cuba	Ubiquitous	Hawaii

Table 3.7. Taxonomy of physaliae

Kingdom	Subkingdom	Branch	Class	Order	Sub-order	Group	Family	Genus	Species
Animal	Coelenterates	Cnidaria	Hydrozoa	Siphono-phora	Sipho-nanta	Physo-phora	Physalida	*Physalia*	*physalis utriculus*

Physalia physalis, commonly known as the sea caravel or Portuguese man-o'-war (in the 15th century English seamen gave it this name because of its resemblance to the caravel, the little ship used by Portuguese seamen to explore the seas) has a main floating nucleus (the pneumatofore, or float) that looks like a large oblong bladder 10-20 cm long, surmounted by a high oblique crest (it is this that gives it the vernacular name of "Portuguese man-o'-war"). The nucleus is filled with gas (a mixture of oxygen, nitrogen and argon), secreted by differentiated cells, whose release is regulated by an orifice with a sphincter at the tip of the float. The portion above the water level is but the tip of the iceberg and from the underside of the float hang the reproductive organs (gonozooids), gastrozooids, dactylozooids and slender tentacles covered with nematocysts. The pneumatophore and the elements below it make up a "colony". The tentacles capture the prey and by retraction and extension movements carry it to the gastrozooids. When these tentacles are fully stretched they may be as fine as strands of human hair and from 10 to 100 foot long (3-30 m). This characteristic, together with their transparency and the fact that they can float far behind the physalia, makes them extremely hazardous. In transparent waters, physaliae look like small, bloated, blue plastic bags but they are in fact very difficult to see, as they blend with the blue water.

The tentacles are studded with nematocysts arranged as a spiral, at the level of the mouth and gastric filaments. Physalia cnidocysts can penetrate the skin of the palms even through rubber gloves. The venom they contain is a protein complex consisting of 8-9 peptides; this is a very labile toxin that is inactivated at 55° C. It has weak antigenic properties, a slight necrotizing action, cardiotoxic activity and a fatal neuromyotoxic activity. The pain after contact with a physalia is due to particular enzymes or substances similar to quinines. The poison is urticant to man and paralysing to its prey (a fish is paralysed in a few seconds): 2 g of fresh filament are enough to induce the death of a 300 g pigeon in one hour, after injection in the great pectoral muscle. The toxin isolated from the venom is called hypnotoxin because of its hypnotic properties; the paralysing action is thought to be due to quaternary ammonium salts [90]. After contact, the poison is injected into the tissues in a fraction of a second and provokes what is commonly called the "physalic syndrome" [1, 5, 46, 90-94].

Skin and systemic reactions

The pain is extremely violent and can rapidly become unbearable, inducing reflex syncope. It radiates out from the affected area and is accompanied by intense burning sensations. The objective picture is characterized by erythemato-oedematous linear lesions; the erythema is bright red at the centre and surrounded by a darker area. Vesicles and blisters can develop on these lesions. Sometimes after a few hours an urticarial eruption can be observed with wide wheals which are particularly itchy.

A few minutes after contact, the victim develops a state of anxiety, anguish and the feeling of imminent death, and then lipothymia. Muscle pain may also be present (with violent curvature of the dorso-lumbar area), asthma-like breathlessness, nausea with or without vomiting, weakness (following a brief phase of excitement and euphoria), bradycardia and hypothermia. In cases of eye involvement, there may be conjunctivitis, intense oedema, painful corneal ulceration and scarring.

In benign cases the skin lesions resolve after a few hours leaving hyperpigmented areas that sometimes persist for months, or scars. The general conditions improve rapidly, while breathlessness and a sharp cough, generalized urticaria and violent digestive symptoms (vomiting, painful colic) may persist for some days. In our latitudes, coma is a rare late complication but it is common in tropical zones [95-97]. Sometimes the linear skin lesions can turn into deep purulent sores.

In a recent study, Burnett and coll. reported a severe case of poisoning by *Physalia physalis* in an underwater diver in the Atlantic [98]. A scuba diving instructor was emerging one evening (7.30 p.m.) without a torch from a depth of 9 m near Miami, Florida. The wetsuit left his face and neck free and he was wearing gloves and carrying a lobster in each hand. As he surfaced, the tentacles of a Portuguese man-o'-war struck him in the face. To free himself, he turned over in the water, which caused the tentacles to wind around his neck and face. Systemic symptoms onset rapidly and his extremely severe respiratory, muscular, intestinal and neurological conditions required long, intensive hospital care. After 5 years, anti-*Physalia* IgG antibodies at titres of 1:450 (normal range 1:50 or less) were isolated in his serum.

The authors pointed out some precautions that could prevent such dramatic episodes: scuba divers must wear all-over wetsuits; during evening immersions they must carry a torch to see the animal from below; during emersion they must look upwards with one arm outstretched towards the surface (even if the tentacles are not seen, they will wind innocuously around the covered arm). When a subject is stung, he/she must resist the natural temptation to break free from the animal, that will only adhere more closely over a wider surface as a result of these manoeuvres. The tentacles must not be removed in the water (manipulation increases their offensive potential) but only once outside.

Reactions to hydroids

Many rocks and the tips of corals are encrusted with an irregular layer of sponges, seaweed, Tunicatae and colonies of small polyps (hydroids) of the Hydrozoa class. Only a few hydroids can be distinguished from the other animals and plants attached to hard surfaces. They form colonies about 5 cm high, some of which resemble white plumes and others slender candelabra. The most common species belong to three families (Sertulariidae, Plumulariidae and Aglaopheniidae), of the Leptomedusa order, and live in tropical and subtropical waters. When they are inadvertently touched, many species induce severe symptoms, especially the feather-like *Lytocarpus philippinus* [3]. Their venom, that acts more slowly than jellyfish venom, can induce two types of reaction: an urticarial eruption after a few minutes from contact, and a haemorrhagic, papulous or zoster-like reaction after 4-12 hours. The skin symptoms are associated with systemic symptoms. The subject may become sensitised.

Reactions to corals

The order of corals (Scleractinia) belongs to the Anthozoa class of the Coelenterate phylum. Corals create an underwater landscape of white rock, studded with thousands of holes. At night, a hand with 6 or more fingers extends out of each hole towards whatever living creature is passing nearby. To complete the bizarre scenario, these "hands" are covered with hairs, and contact with these can be very harmful. Each "hand" is connected to the others by a narrow strip of tissue that stretches over the surface of the rock, so that the hole is nothing other than a sheath of living tissue. The "hands", known as polyps or animal part of the coral, are covered with cnidoblasts and the fingers are retractile (Fig. 3.47).

There are two different types of coral, classified according to whether they have a hard or soft skeleton. The polyps with a hard shell deposit a solid skeleton of calcium carbonate around themselves, secreted by the epidermal cells: in this way, the animal builds itself a cup-like shell (fossil coral) where it lives during the day. The skeleton is the only visible part of the coral, at least by day, and it is the skeleton which is left when the animal dies [3, 99-101].

Fig. 3.47. *Corallium rubrum* and *Parazoanthus axillinae* (sea anemone)

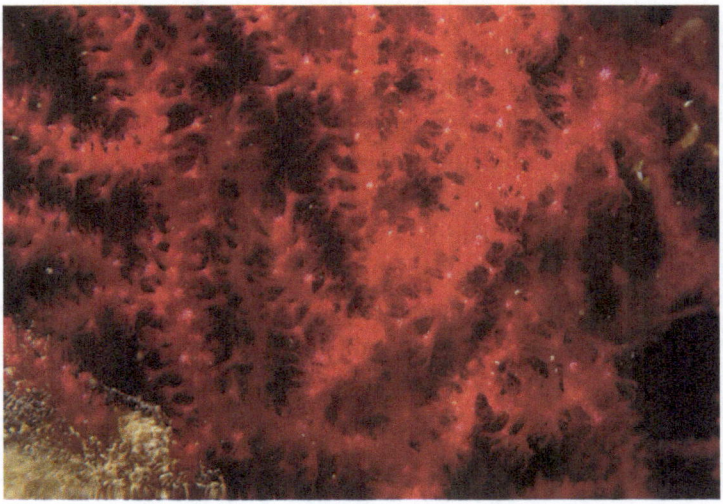

Fig. 3.48. *Paramuricea clavata* (red coral)

Soft corals, of the Alcyonaria subclass, secrete the same kind of calcium carbonate skeleton: the difference lies in the internal cohesion and microcrystalline structure, which determine the consistency of the coral. These corals, named Gorgonia (because they have a flexible axial skeleton) (Fig. 3.48) are abundant in the Caribbean sea, for some unknown reason.

Corals can provoke skin lesions of various types. Toxic contact reactions are relatively infrequent and generally fairly mild, rather like those induced by jellyfish and sea anemones. Probably, however, the frequency of irritant contact reactions is underestimated [101].

In comparison with the toxins in coral nematocysts that are not very harmful to man, those of "stinging or fire corals" of the *Millepora* genus (Milleporina order of the Hydrozoa class) are much more serious. These hydroid corals form a hard skeleton that appears very similar to that of true corals; they are very widespread in tropical seas in shallow waters. They give rise to cleaved and branching calcareous formations that can encrust other corals and objects and range in colour from white to yellowish-green. The best known species are *Millepora alcicornis, M. complanata* and *M. squarrosa.*

The main clinical manifestations attributed to their nematocysts are: erythema with associated itching, contact urticaria, eczema, vesico-bullous eruptions and lichenoid and granulomatous lesions. Sometimes several different clinical pictures can be observed in succession in the same patient. For instance, some cases have been described where the dermatitis was of urticarial type at the time of contact, which then rapidly evolved into erythemato-oedemato-bullous lesions, followed by persistent lichenoid manifestations [102]. In other cases, delayed reactions have featured generalized lichenoid or granulomatous lesions that onset a few weeks after the primitive contact [103]. Observations of persistent and recurrent contact dermatitis have led some researchers to assume that there is not only a type I hypersensitivity response but also a cell-mediated immune response, which has been confirmed by histological and immunohistochemical findings [103, 104]. In exceptional cases, generalized symptoms such as fever and nausea can be observed.

Instead, wounds from corals are very frequent: despite their fragile appearance, hard corals have very sharp, cutting surfaces. These wounds rapidly evolve into painful ulcers and unless they are promptly and appropriately treated, into cellulitis. The severity of the latter picture is due to a combination of a number of different factors: mechanical skin laceration, the offensive action of the nematocysts, the introduction of foreign bodies into the wound (calcium carbonate, detritus, micro-organisms, sludge), secondary bacterial infections and climatic conditions (high temperature and humidity) favouring the development of bacteria. These wounds heal very slowly.

Diagnosis, prognosis and treatment

The correct treatment of Coelenterate poisoning syndromes obviously requires a correct clinical diagnosis and the identification of the species responsible [105]. In general, the clinical diagnosis is not particularly difficult, although the possible onset of skin and systemic reactions even without direct contact with a Coelenterate should be kept in mind.

To identify subjects at risk, a RAST has been developed for Coelenterate toxins [30, 106-107]. Some studies have shown that repeated exposure to these animals can result in the formation of specific IgE; that significant levels of these circulating antibodies can persist for several years; and that antibodies to one species can cross-react with antigens from other species.

In affected patients, specific IgG-blocking antibodies have also been demonstrated with RAST using protein A of *Staphylococcus aureus* to bind the IgG [107]. The presence of the latter type of antibodies may have a protective effect, while the increased specific IgE levels in the absence of blocking antibodies suggest a particular susceptibility. For this reason, before "labelling" a patient with specific IgE levels as a subject at risk, it is best to ascertain the IgG-blocking levels, too.

There are no safe preventive measures against Coelenterate poisoning. In Australia and the USA, where this is a real problem as there is a risk of fatal outcome, special protective wetsuits are produced that have been shown to answer the purpose, or at any rate to reduce the severity of the damage. Instead, the use of barrier creams (that obviously do not protect the eyes) and mechanical gratings closing off outside access to some Australian beaches have not been found to be effective [7].

Until all the toxins of the various Coelenterate species have been identified, the treatment will necessarily be aspecific and symptomatic. At present only one specific antiserum is available, for poisoning by *Chironex fleckeri*, that has also been found to be useful against *Chiropsalmus quadrigatus*; this is produced by the Commonwealth Serum Laboratories of Melbourne and must be inoculated as soon as possible after contact with the jellyfish.

To alleviate the symptoms of Coelenterate stings, fishermen and the inhabitants of seaside towns use old-wives' remedies that still have some uses, such as vinegar (5% acetic acid in water), ammonia, urine, formaldehyde, potassium permanganate crystals, warm water, Coca-Cola [108] and ice.

The following measures should be borne in mind when treating these conditions [1, 5, 7, 48, 90, 109]. It is absolutely essential to avoid using freshwater as this is hypotonic and can cause the nematocysts to burst. For the same reason, the subject must not shower until the toxins have been neutralized. Instead, the affected skin areas can be gently washed with seawater without any risk.

As to the use of alcohol, *in vitro* it has been observed to stimulate nematocyst bursting, although many experiences have described this means as useful. Hydroalcoholic solutions, such as perfumes, after-shave lotions and spirits (ethanol) must not be used, as in some cases they prolong the agony. As an alternative to vinegar and ammonia, the proteolytic enzymes of meat (papaine) can be used, since they have the same action. If all else is lacking, saltwater heated to the limit of the patient's tolerance contributes to neutralize the poison.

The skin must not be rubbed to remove sand, again to prevent the nematocysts from bursting. The use of a formalin solution can fix the nematocysts and stop them from bursting more effectively than ammonia and vinegar. Sodium bicarbonate and alkaline solutions can also neutralize the toxins; in some cases even sun lotions have been found to help.

To remove the tentacles, thick gloves must be used, or a paste made with saltwater and sodium bicarbonate left on for 5 minutes. The use of talcum powder or flour can have the same effect, causing agglomeration of the tentacles that can then be removed with a knife or sharp tool. Instead of powder, dry sand can be used, after which the zone must again be carefully washed with seawater.

Topical treatment is based on corticosteroids and anaesthetics in the form of creams, lotions or aerosols, to relieve itching and burning sensations. It is wise to avoid products with a benzocaine base, as this has a potential sensitising action, and to select lidocaine derivatives at 5%. As topical products do not penetrate rapidly, systemic anaesthetics are preferable. Instead, when the eyes are affected these remedies are not indicated as they could cause further harm, and only topical steroids should be used.

In particularly severe cases, especially in children, a haemostatic rubber tourniquet can be used if a limb has been affected. This aims to impede venous return, not to block arterial flow and the tourniquet must be loose enough for one finger to be able to pass under it and must be undone for 3-4 minutes every hour.

For systemic treatment, antihistamines and corticosteroids are very useful, as well as cardiotonics. States of shock must be treated with epinephrine, together with systemic corticosteroids. Cramps can be treated by intravenous calcium gluconate, while a combination of aspirin, codeine and phenacetin helps to alleviate acute pain.

Necrotic and ulcerative lesions must be cleansed 3 times a day and treated with topical antibiotics (erythromycin, tetracycline). These antibiotics can also be used systemically in cases of secondary infections or involvement of vast skin areas.

The doctor and first aid workers must be aware that nematocysts that have been detached from the tentacles can maintain their toxic action for several months. For this reason, bathers must avoid areas infested with Coelenterates, especially after a storm, that can favour the spread of tentacle segments in the water and to the shore.

Finally, sequelae such as hyperchromia can be treated with topical bleaches and cheloids with the conventional methods.

References

1. Fisher AA (1978) Atlas of aquatic dermatology. Grune and Stratton, New York
2. Altamura BM, Introna F, Rositani L (1981) Lesività da fauna marina mediterranea. Med Leg 3:13
3. Kaplan EH (1982) Coral reefs. Peterson Field Guides. Houghton Mifflin Company, Boston, 55

4. Ghiretti F, Cariello L (1984) Gli animali marini velenosi e le loro tossine. Piccin, Padova
5. Fisher AA (1986) Aquatic dermatitis. In: Fisher AA (ed) Contact dermatitis, 3rd edn. Lea and Febiger, Philadelphia, 809
6. Angelini G, Vena GA (1991) Principi di dermatologia acquatica. Dermotime 3:15
7. Burnett JW (1992) Human injuries following jellyfish stings. Mol Med J 46:509
8. Halstead BW (1992) Dangerous aquatic animals of the world: a color atlas. The Darwin Press Inc, Princeton, 31
9. Gowell ET (1993) Sea jellies. Rainbows in the sea. Franklin Watts, New York
10. Angelini G, Bonamonte D (1997) Dermatoses aquatiques méditerranéennes. Nouv Dermatol 16:280
11. Angelini G, Vena GA (1997) Dermatologia professionale e ambientale. Vol I. ISED, Brescia
12. Angelini G (2000) Occupational aquatic dermatology. In: Kanerva L, Elsner P, Wahlberg JE et al (eds) Handbook of occupational dermatology. Springer Berlin Heidelberg New York, 234
13. Kokelj F (2000) Patologia da meduse. In: Veraldi S, Caputo R (eds) Dermatologia di importazione. Poletto Editore, Milano, 286
14. Foti C, Bonamonte D, Vena GA et al (2000) Dermatiti da attinie. In: Veraldi S, Caputo R (eds) Dermatologia di importazione. Poletto Editore, Milano, 297
15. Mariscal RN (1974) Nematocysts. In: Muscatine L, Lenhoff HM (eds) Coelenterate biology. Academic Press, New York, 129
16. Tardent P, Honegger T, Baenninger R (1980) About the function of stenotheles in *Hydra attenuata Pall*. In: Tardent P, Tardent R (eds) Developmental and cellular biology of Coelenterates. Elsevier, Amsterdam
17. Tardent P, Holstein T (1982) Morphology and morphodynamics of the stenothele nematocyst of *Hydra attenuata Pa.* (Hydrozoa, Cnidaria). Cell Tissue Res 224:269
18. Holstein T, Tardent T (1984) An ultralight-speed analysis of exocytosis: nematocyst discharge. Science 223:830
19. Tardent P (1995) The cnidarian cnidocyte, a high-tech cellular weaponry. Bioessays 17:351
20. Portier P, Richet C (1902) Sur les effects physiologiques du poison des filaments pêcheurs et des tentacules des Coelentérés (hypnotoxine). C R Acad Sci III 134:247
21. Richet C (1903) Des poisons contenus dans les tentacules des actinies (congestine et thalassine). C R Soc Biol 55:246
22. Richet C (1903) De la thalassine, toxine cristallisée pruritogène. C R Soc Biol 55:707
23. Arillo A, Burlando B, Carli AM et al (1994) Mitochondrial alteration caused by cnidarian toxins: a preliminary study. Boll Soc Ital Biol Sper 70:307
24. Chàvez M, Gil S, Fernandez A et al (1998) Purification and partial characterization of a proteinase inhibitor from sea anemone *Condylactis gigantea*. Toxicon 36:1275
25. Diaz J, Morea V, Delfin J et al (1998) Purification and partial characterization of a novel proteinase inhibitor from the sea anemone *Stichodactyla helianthus*. Toxicon 36:1275
26. Aneiros A, Karlsson E, Beress L et al (1998) Isolation of toxins from the Caribbean sea anemones *Bunodosoma granulifera* and *Phyllactis floscuifera*. Toxicon 36:1276
27. Moore RE, Scheuer PJ (1971) Palytoxin: a new marine toxin from a Coelenterate. Science 172:495
28. Uemura D, Ueda K, Hirata Y et al (1981) Further studies on palytoxin. II. Structure of palytoxin. Tetrahedron Lett 22:2781
29. Kokelj F (1996) Jellyfish stinging in the Mediterranean Sea. In: Williamson JA, Fenner PJ, Burnett JW (eds) Venomous and poisonous marine animals: a medical and biological handbook. University of New South Wales Press, Sidney
30. Kokelj F, Burnett JW (1988) Reazioni inusuali indotte dal contatto con la medusa *Pelagia noctiluca*. Presentazione di tre casi. G Ital Dermatol Venereol 123:501
31. Burnett JW, Cobbs CS, Kelman SN et al (1983) Studies on the serologic response to jellyfish envenomation. J Am Acad Dermatol 9:229

32. Togias AG, Burnett JW, Kagei-Sobotka A et al (1985) Anaphylaxis after contact with a jellyfish. J Allergy Clin Immunol 75:672

33. Michaeli D, Benjamini E, Miner RC et al (1966) In vitro studies on the role of collagen in the induction of hypersensitivity to flea bites. J Immunol 96:402

34. Russo AJ, Calton GJ, Burnett JW (1983) The relationship of the possible allergic response to jellyfish envenomation and serum antibody titers. Toxicon 21:475

35. Mansson T, Randle HW, Mandojana RM et al (1985) Recurrent cutaneous jellyfish eruption without envenomation. Acta Derm Venereol 65:72

36. Reed KM, Bronstein BR, Baden HP (1984) Delayed and persistent cutaneous reactions to Coelenterates. J Am Acad Dermatol 10:462

37. Burnett JW, Hepper KP, Aurelian L et al (1987) Recurrent eruptions following unusual solitary Coelenterate envenomations. J Am Acad Dermatol 17:86

38. Russel FS (1970) The medusae of the British Isles. Cambridge University Press, Cambridge

39. Kokelj F, Del Negro P, Tubaro A (1989) Dermotossicità da *Chrysaora hysoscella*. Presentazione di un caso. G Ital Derm Venereol 124:297

40. Del Negro P, Kokelj F, Avian M et al (1991) Toxic property of the jellyfish *Chrysaora hysoscella*: preliminary report. Rev Intern Océanograph Méd 101:168

41. Kokelj F, Del Negro P, Montanari G (1992) Jellyfish dermatitis due to *Carybdaea marsupialis*. Contact Dermatitis 27:195

42. Kokelj P, Avian M, Spanier E et al (1995) Dermatotoxicity of 2 nematocyst preparations of the jellyfish *Rhopilema nomadica*. Contact Dermatitis 32:244

43. Long-Rowe KO, Burnett JW (1994) Characteristics of hyaluronidase and hemolytic activity in fishing tentacle nematocyst venom of *Chrysaora quinquecirrha*. Toxicon 32:165

44. Long-Rowe KO, Burnett JW (1994) Sea nettle (*Chrysaora quinquecirrha*) lethal factor: purification by recycling on *m*-aminophenyl boronic acid acrylic beads. Toxicon 32:467

45. Houck HE, Lipsky MM, Marzella L et al (1996) Toxicity of sea nettle (*Chrysaora quinquecirrha*) fishing tentacle nematocyst venom in cultured rot hepatocytes. Toxicon 34:771

46. Burnett JW, Calton GJ, Burnett HW (1986) Jellyfish envenomation syndromes. J Am Acad Dermatol 14:100

47. Rosco MD (1977) Cutaneous manifestations of marine animal injuries including diagnosis and treatment. Cutis 19:507

48. Burnett JW, Calton GJ, Morgan RJ (1987) Venomous Coelenterates. Cutis 39:191

49. Fisher AA (1987) Toxic and allergic cutaneous reactions to jellyfish with special reference to delayed reactions. Cutis 40:303

50. Burnett JW, Calton GJ (1977) The chemistry and toxicology of some venomous pelagic Coelenterates. Toxicon 15:177

51. Glasser DB, Noell MJ, Burnett JW et al (1992) Ocular jellyfish stings. Ophthalmology 99:1414

52. Peters H (1967) Hydrodein-dermatitis. Hautarzt 18:396

53. Matusow RJ (1980) Oral inflammatory responses to a sting from a Portuguese man-of-war. J Am Dent Assoc 100:73

54. Kromp P (1961) Synopsis of the medusae of the world. J Mar Biol Assoc UK 40:1

55. Kokelj F, Mianzan H, Avian M et al (1993) Dermatitis due to *Olindias sambaquiensis*: a case report. Cutis 51:339

56. Kokelj F, Stinco G, Avian M et al (1995) Cell-mediated sensitization to jellyfish antigens confirmed by positive patch test to *Olindias sambaquiensis* preparations. J Am Acad Dermatol 33:307

57. Yaffee HS (1968) A delayed cutaneous reaction following contact with jellyfish. Dermatol Int April-June issue, p 75

58. Williamson JA, Le Ray LE, Wohlfart M et al (1984) The acute management of serious box-jellyfish (*Chironex fleckeri*) stings. Med J Aust 141:851

59. Querull P, Bernard P, Dantzer E (1996) Severe cutaneous envenomation by the Mediterranean jellyfish *Pelagia noctiluca*. Vet Hemsan Toxicol 38:460

60. Gunn MA (1947) Localized fat atrophy after jellyfish sting. Br Med J 2:687
61. Drury JK, Noonan JH, Pollock JH et al (1980) Jellyfish sting with serious hand complications. Injury 12:66
62. Williamson JA, Burnett JW, Fenner PJ et al (1988) Acute regional vascular insufficiency after jellyfish envenomation. Med J Aust 149:697
63. Burnett JW, Williamson JA, Fenner PJ (1994) Mononeuritis multiplex after Coelenterate sting. Med J Aust 161:320
64. Filling-Katz MR (1984) Mononeuritis multiplex following jellyfish stings. Ann Neurol 15:213
65. Moats WE (1992) Fire coral envenomation. J Wilderness Med 3:284
66. Peel N, Kandler R (1990) Localized neuropathy following jellyfish sting. Postgrad Med J 66:953
67. Flecker H (1952) Irukandji sting to north Queensland bathers without production of weals but with severe general symptoms. Med J Aust 2:89
68. Southcott RV (1967) Revision of some Carybdeidae (Schyphozoa: cubomedusae) including a description of the jellyfish responsible for the "Irukandji syndrome". Aust J Zool 15:651
69. Little M, Mulcahy RF (1998) Bites and sting: a year's experience of Irukandji envenomation in far north Queensland. Med J Aust 169:638
70. Wiltshire CJ, Sutherland SK, Winkel KD et al (1998) Comparative studies on venom extracts from three jellyfishes: the Irukandji (*Carukia barnesi*), the box jellyfish (*Chironex fleckeri southcott*) and the blubber (*Catosylus mosaicus*). Toxicon 36:1239
71. Burnett JW, Calton GJ, Fenner PJ et al (1988) Serological diagnosis to jellyfish envenomations. Comp Biochem Physiol 91C:79
72. Wachsman M, Aurelian L, Burnett JW (1991) Human immunosuppression induced by sea nettle (*Chrysaora quinquecirrha*) venom. Toxicon 29:386
73. Burnett JW, Bloom DA, Imafuku S et al (1996) Coelenterate research 1991-1995: clinical, chemical and immunological aspects. Toxicon 34:1377
74. Ferlan L, Lebez D (1974) Equinatoxin, a lethal protein from *Actinia equina*. I. Purification and characterization. Toxicon 12:57
75. Maretec Z, Russel FE (1963) Stings by the sea anemone *Anemonia sulcata* in the Adriatic sea. J Trop Med Hyg 32:891
76. Vena GA, Fiordalisi F, Angelini G (1989) Dermatite da contatto e reazione anafilattoide da *Anemonia sulcata*. In: Ayala F, Balato N (eds) Dermatologia in Posters. Cilag S.p.A., Napoli
77. Molfino F, Zannini D (1964) L'uomo e il mondo sommerso. Medicina subacquea. Minerva Med, Torino
78. Sams WM (1949) Seabather's eruption. Arch Dermatol 60:227
79. Pike AW (1989) Sea lice: major pathogens of farmed Atlantic salmon. Parasitol Today 5:291
80. Hutton RF (1960) Marine dermatosis. Arch Dermatol 82:951
81. Tomchik RS, Russel MT, Szmant AM et al (1993) Clinical perspectives on seabather's eruption, also known as sea lice. JAMA 269:1669
82. Straus JS (1956) Seabather's eruption. Arch Dermatol 74:293
83. Moschella H (1951) Further clinical observations on seabather's eruption. Arch Dermatol 64:55
84. Frankel EH (1992) Seabather's eruption develops following Mexican vacation. Clin Cases Dermatol 4:6
85. Freudenthal AR (1991) Seabather's eruption: range extended northward and a causative organism identified. Rev Int Oceanogr Med 101:137
86. Wong DE, Meinking TL, Rosen LB et al (1994) Seabather's eruption. Clinical, histologic, and immunologic features. J Am Acad Dermatol 30:399
87. Angelini G, Vena GA (2001) Dermatosi acquageniche. In: Giannetti A (ed) Trattato di dermatologia. Vol II. Piccin, Padova, cap. 45
88. Russell MT, Tomchik RS (1993) Seabather's eruption, or "sea lice": new findings and clinical implications. J Emerg Nurs 93:197

89. Jefferies NJ, Rushby N (1997) Caribbean itch: eight cases and one who didn't. J Royal Army Med Corps 143:163
90. Ducombs G, Lamy M (1985) Accidents dus à *Physalia physalis* L. " Le syndrome physalien ". Bull Act Thérap 30:3011
91. Ioannides G, Davis JH (1965) Portuguese man-of-war stinging. Arch Dermatol 91:448
92. Marr JJ (1967) Portuguese man-of-war. Envenomization. A personal experience. JAMA 199:337
93. Russel FE (1966) *Physalia* stings: a report of two cases. Toxicon 4:65
94. Baslow MH (1969) Marine pharmacology. Williams and Wilkins Co, Baltimore
95. Burnett JW, Gable WD (1989) A fatal jellyfish envenomation by the Portuguese man-of-war. Toxicon 27:823
96. Stein MR, Marraccini JV, Rothschild NE et al (1989) Fatal Portuguese man-of-war (*Physalia physalis*) envenomation. Ann Emerg Med 18:312
97. Bonamonte D, Cassano N, Angelini G et al (2000) Dermatiti da fisalie e da idroidi. In: Veraldi S, Caputo R (eds). Dermatologia di importazione. Poletto Editore, Milano, 309
98. Burnett JW, Fenner PJ, Kokelj F et al (1994) Serious *Physalia* (Portuguese man-of-war) stings: implications for scuba divers. J Wilderness Med 5:71
99. Banister K, Campbell A (1993) The encyclopedia of aquatic life. Facts on File, New York
100. Bonamonte D, Foti C, Vena GA et al (2000) Dermatiti da coralli. In: Veraldi S, Caputo R (eds). Dermatologia di importazione. Poletto Editore, Milano, 311
101. Tong D (1995) Coral dermatitis in the aquarium industry. Contact Dermatitis 33:207
102. Addy JH (1991) Red sea coral contact dermatitis. Int J Dermatol 30:271
103. Camarasa JG, Nogués Antich E, Serra-Baldrich E (1993) Red sea coral contact dermatitis. Contact Dermatitis 29:285
104. Piérard GE, Letot B, Piérard-Franchimont C (1990) Histologic study of delayed reactions to Coelenterates. J Am Acad Dermatol 22:599
105. Fenner PJ, Williamson JA, Burnett JW (1998) Treatment and prevention of jellyfish envenomation. Toxicon 36:1242
106. Hartman KR, Calton GJ, Burnett JW (1980) The use of the radio-allergosorbent test for the study of Coelenterate toxin-specific immunoglobulin E. Int Arch Allergy Immunol 61:389
107. Burnett JW, Calton GJ (1981) Use of IgE antibody determinations in cutaneous Coelenterate envenomations. Cutis 27:50
108. Currie B, Stephen H, Ho S (1993) Box-jellyfish, Coca-Cola and old wine. Med J Aust 158:868
109. Afa G, Olivetti G (1986) Dermatologia acquatica mediterranea. La Lettera del Dermatologo 5:4

4 Dermatitis caused by Echinoderms

Fig. 4.1. *Sphaerechinus granularis* and *Echinaster sepositus*

Echinoderms (*echinos* = bristle or spiny appearance) (Echinodermata phylum) are animals with a rotate pentamerous symmetry. About 6,000 species are known, 80 of which are toxic or venomous. Their various different shapes have led to a subdivision into five classes. Some are spherical and covered in spines or strong spicules (Echinoidea or sea urchins); some are star-shaped with five points or ray-like arms of variable length (Asteroidea or starfish) (Fig. 4.1); some have a cylindrical body (Holothuroidea or sea cucumbers) (Fig. 4.2); some are flower-like (Crinoidea, sea lilies or feather stars); others have long, branching arms that can twine around solid bodies (Ophiuroidea) (serpent stars: *ophis + idis* = snake-like) (Table 4.1) (Fig. 4.3) [1].

Fig. 4.2. Holothuria or sea cucumber

Table 4.1. Echinodermata classes and most common toxic species

1. Echinoidea (sea urchins)	3. Holothuroidea (sea cucumbers)

Species:
  Paracentrotus lividus
  Arbacia lixula
  Sphaerechinus granularis
	Species:
	  Cucumaria
	  Stichopus

2. Asteroidea (starfish)	4. Crinoidea (sea lilies)

Species:
  Echinaster sepositus
  Acanthaster planci
	5. Ophiuroidea (sea snakes)

Fig. 4.3. Crinoidea (sea lily)

Apart from the Holothuroidea, that have a soft body, all the other Echinoderms have a brittle endoskeleton of regularly arranged calcareous plates. The sea urchin skeleton is rigid and inflexible, whereas that of starfish consists of calcareous plates not too close together so that it can move its arms to capture prey or turn over when it falls on its back. Instead, the Crinoidea and Ophiuroidea are much more mobile, and have sinuous swimming movements.

Echinoderms have a water vascular system with tubular cavities for the water to circulate through, and tubular feet extending out from the body. In starfish and sea urchins, these tube feet cover the whole body; they are like narrow tubes fitted with a suction cap at the end. As they can retract and extend, they allow the animal to move slowly along the seabed (this is why they are called pedicellariae) and to capture prey.

Starfish and sea urchins are macrophages and capture their prey while the other Echinoderms are microphages and feed on plankton and detritus. These species can be found in all the seas, from the tropics to the Arctic.

Of the 5 groups, two kinds of starfish and various kinds of sea urchins are poisonous. Some Holothuroidea are poisonous to eat and because they discharge very slimy filaments containing toxins into the water from the visceral extremity, they are poisonous to other fish. The toxin contained (holothurin) is a glucoside; through hydrolysis it gives rise to an aglycon of steroid type and various glucide residues that are highly toxic to all organisms, from protozoa to mammals. The effect is often fatal, being of neurotoxic type and provoking an irreversible blockage of transmission of the nerve impulses at the level of the neuromuscular synapses; there is no known specific treatment [1-6].

Dermatitis from sea urchins

About 750 species of Echinoidea have been identified, some of which are present in the Mediterranean: in particular, *Paracentrotus lividus, Arbacia lixula, Sphaerechinus granularis.*

Sea urchins live on rocks and in crevices underwater and are rarely seen on the sand. Their sharp calcareous spines are mobile and attached to the body by dovetail-like articulations on the test (hard shell). They form by means of calcification of cylindrical projections of the sub-epidermal connective tissue. The calcareous crystals are arranged radially around a central canal containing connective tissue in the post-larval stage, that then calcifies.

The spines contain calcium sulphate, magnesium carbonate, calcium carbonate and silicium. Their surface is coated with a thin organic sheath consisting of pigment, epidermal residues and perhaps glands with a toxic content. Toxic species of sea urchins are to be found along the coasts of Europe, the Atlantic and the Pacific. The toxins are contained not only in the spines but also in the pedicellariae. In some species, toxins are also present in the gonads during the reproductive period. The toxins serve for defence and offence, to capture prey (Echinoidea, like Asteroidea, are macrophages), although the functional significance of the toxins in the gonads is unknown. Sea urchin toxins have not yet been identified but surprisingly, they do not seem to be saponins like those of the Asteroidea and Holothuroidea, even if they have the same neurotoxic action. The toxins in the gonads can be harmful to man when these species are eaten as seafood [7]. However, only rarely have fatal cases been reported; the symptoms generally include nausea, vomiting, diarrhoea, headache and allergic reactions that subside after a short time without any further complications.

The sea urchins present in the Mediterranean belong to the less toxic varieties. Two different kinds of reactions to sea urchins can onset (Fig. 4.4) [8-12].

Immediate reactions

Swimmers for professional (fishermen) and sports purposes must beware of sea urchin spines. These develop in various ways in the different species, and are sharp and very fragile. On contact, they penetrate the skin very easily and break off, making the fragments inside the wound very difficult to extract. Penetration of the spines causes immediate, sharp, burning pain which may last a few hours, followed by skin redness and oedema of the affected part (Fig. 4.5); in some cases the lesions bleed copiously. Torpor, breathlessness and localized muscular pain have also

Fig. 4.4. A sea urchin hawker

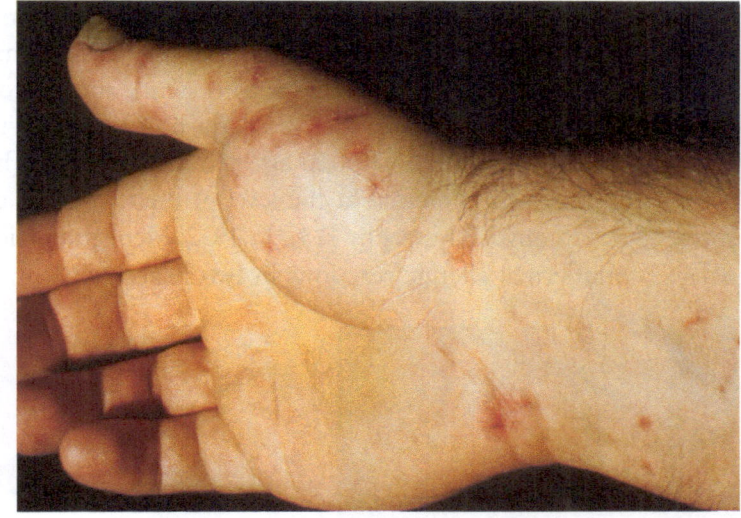

Fig. 4.5. Abrasions from sea urchin spine pricks. Reproduced with permission from [6]

been reported. Secondary infections are not rare, and can eventually cause ejection of the spines.

The treatment of immediate reactions is by applying water as hot as is bearable on the oedematous, painful lesions. Infected lesions must be treated with topical antibiotics. The dermatitis usually resolves within 1-2 weeks, provided no spines are left in the skin. Immediate and complete removal of the spines is obviously a priority, although this is often difficult owing to their fragility and the presence of dyes that make them difficult to distinguish from the surrounding tissue. It should be borne in mind that whilst the spines of some species can be phagocytised by the tissues in 24-48 hours, those of other species do not dissolve. In the latter case surgical removal may be necessary, after an X-ray. Particular complications can arise when the spines enter the area of a joint or in contact with a nerve. The spines of some sea urchins (*Diadema antillarum*) can pierce the soles of shoes, gloves and the wetsuit.

In the West Indies, the spines are extracted using a method that combines scientific and mystical elements. The affected skin part is rubbed with lime juice (citric acid) and then very hot vinegar (acetic acid) is applied. After about 10 minutes, more vinegar is applied and the zone is heated with a candle held at 1 cm from the skin. This causes the spines to dissolve. After a further 10 minutes, candle wax is spread on the part and once this has hardened, it is peeled off bringing the remaining spines with it [4].

Delayed reactions

Delayed reactions can onset 2-3 months after the primitive contact and may be nodular or scleroedematous. Both types of reaction can persist for a very long time, although they may resolve spontaneously. Granulomatous nodular lesions have a hard, parenchymatous consistency and range in size from 4-5 mm to 1-2 cm in diameter; they are darkish or brownish-red in colour (Figs. 4.6-4.11). The sites most often affected are the hands (back, palms), elbows and knees. Granulomas at the level of the nail quick can bring on severe forms of onychodystrophy [13]. In 1972, Meneghini elicited positive delayed intradermal allergic reactions in two fishermen with granulomas,

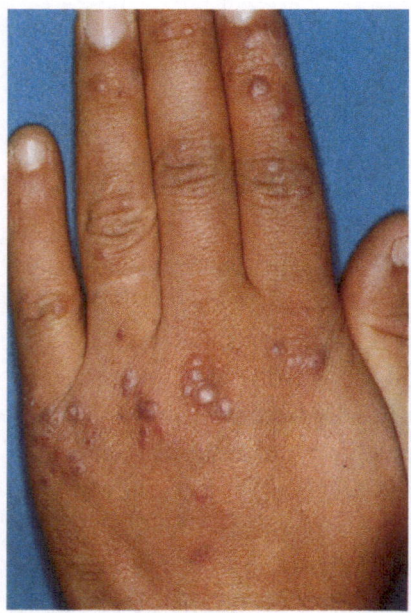

Fig. 4.6. Granulomas from sea urchins. Reproduced with permission from [6]

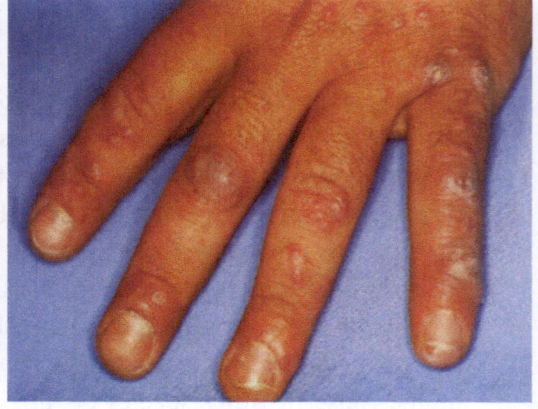

Fig. 4.7. Granulomas from sea urchins

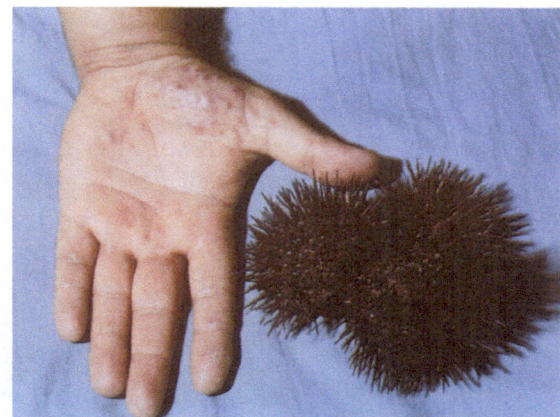

Fig.   4.8. Granulomas   from *Paracentrotus lividus* (edible sea urchin). Reproduced with permission from [8]

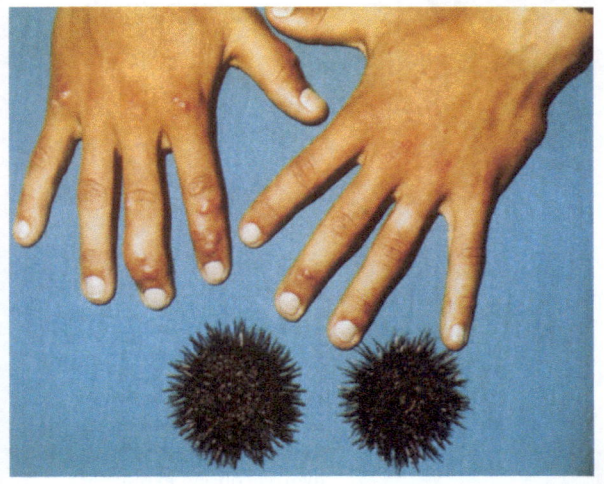

Fig. 4.9. Granulomas from *Paracentrotus lividus* (edible sea urchin). Reproduced with permission from [6]

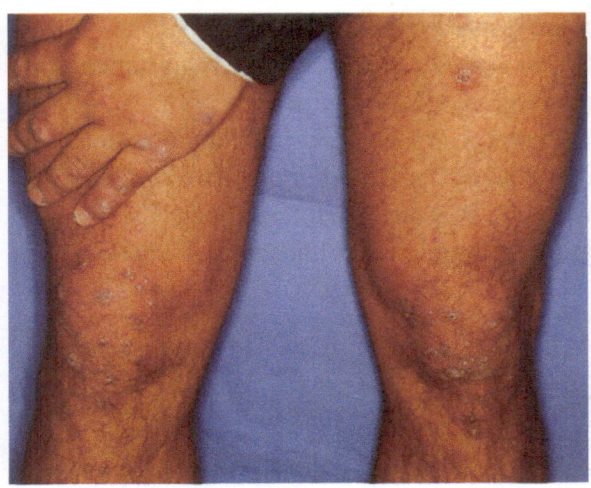

Fig. 4.10. Granulomas from sea urchins on typical sites

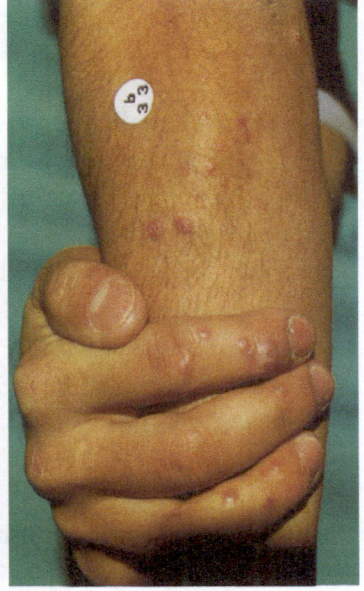

Fig.   4.11. Granulomas   from   sea urchins on typical sites

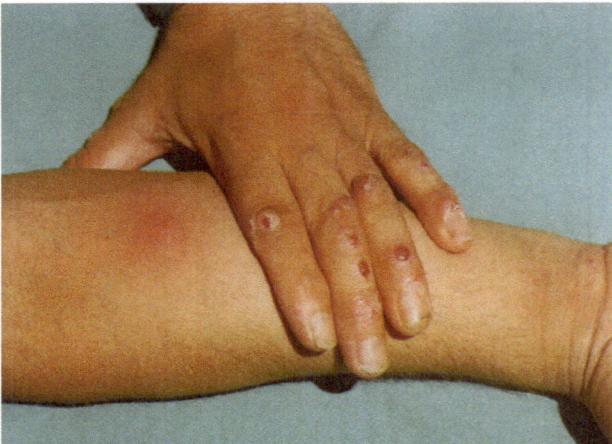

Fig. 4.12. Positive delayed reaction after 48 hours to intradermal test with hydroalcoholic extract of sea urchin spines. Reproduced with permission from [6]

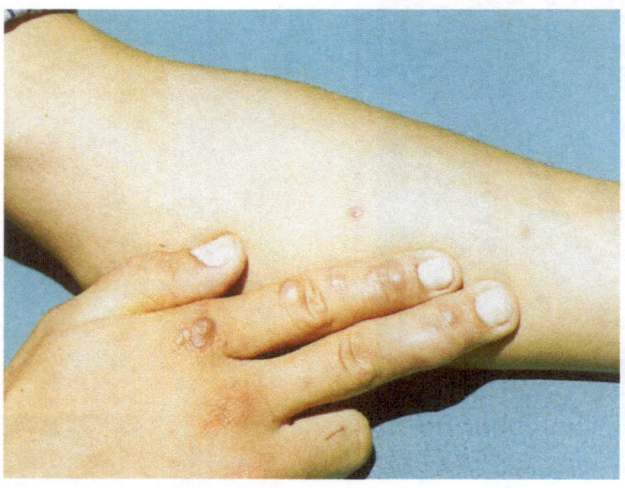

Fig. 4.13. Positive delayed reaction after 48 hours to intradermal test with hydroalcoholic extract of sea urchin spines

using hydroalcoholic extracts of sea urchin spines [11]. We have since observed the same reactions in various other subjects (Figs. 4.12, 4.13).

When they penetrate the skin, the spines can transfer other substances such as seaweed, that can also induce granulomatous reactions [14].

Various types of histological findings are possible in granulomatous lesions: microabscesses, sarcoid-type granulomas [10], or more often lympho-histiocytic granulomatous foci with giant cells due to a foreign body [15].

Nodular lesions can be treated with intralesional injections of cortico-steroids or liquid nitrogen.

Occupational chronic traumatic scleroedema

Diffuse delayed reactions of the backs of the hands may manifest as a particular form of chronic scleroedema. This traumatic lymphoedema is a pro-

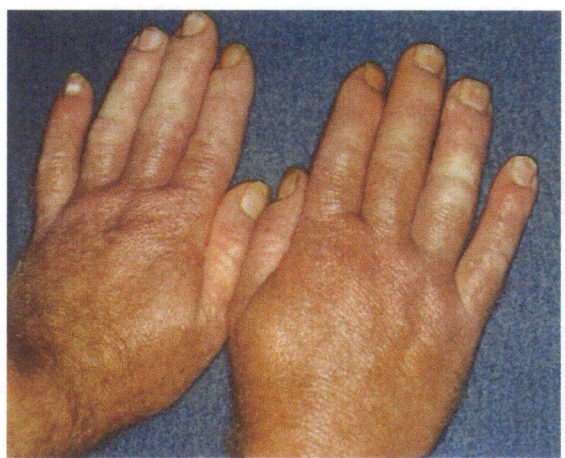

Fig. 4.14. Chronic traumatic scleroedema of the hands in a sea urchin fisherman. Acrocyanosis and skin atrophy are also evident

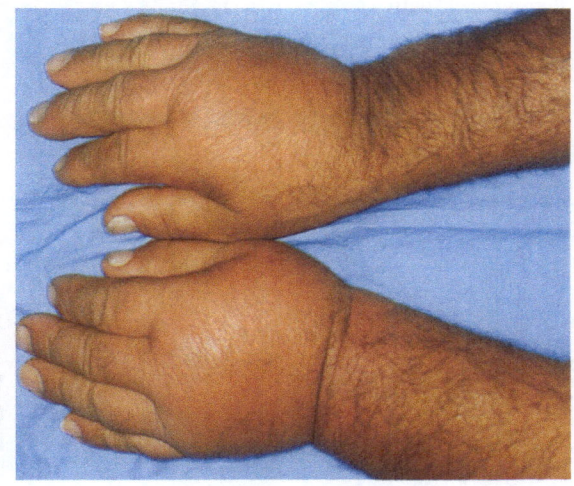

Fig. 4.15. Intense chronic traumatic scleroedema of the hands and forearms. The fisherman had stopped this working activity 15 years before. The joint function is impaired. Reproduced with permission from [6]

fessional complaint and we have often observed it in fishermen, caused by repeated penetration of sea urchin spines, together with the constriction of the wrists caused by the wetsuit and the low temperature of the water. It manifests with hard, persistent oedema of the backs of the hands and sometimes also of the forearms (Figs. 4.14, 4.15) [16-18].

The oedema is firstly recurrent but within a few years it becomes persistent, very hard and clearly distinct, ending in a sharp line at the wrists. It can persist for many years even after abandonment of this working activity and may be associated with "sea urchin granulomas" (Figs. 4.16, 4.17), functional impairment of the wrists and fingers, dystrophic alterations of the nails and sometimes acrocyanosis, "cigarette-paper" atrophy of the affected skin and joint dysmorphisms. In one case with intense scleroedema and granulomas, lymphography of the upper limb showed irregular spread and distribution of the contrast medium on the back of the hand (Fig. 4.18) [17].

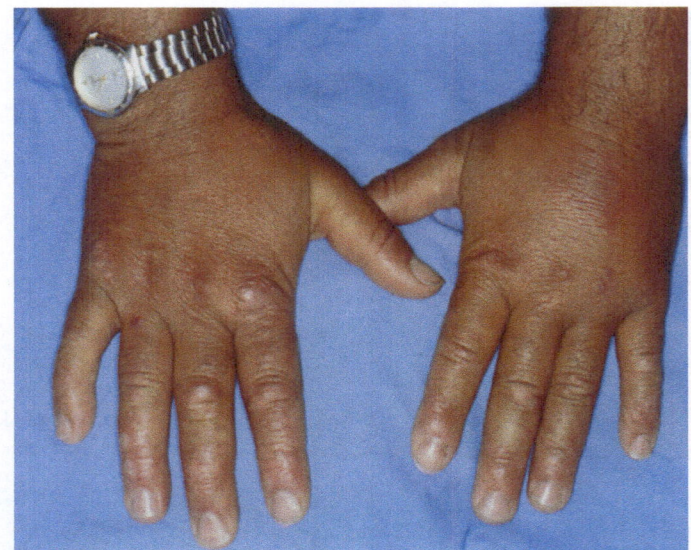

Fig. 4.16. Chronic traumatic scleroedema of the hands and granulomas from sea urchins. Reproduced with permission from [8]

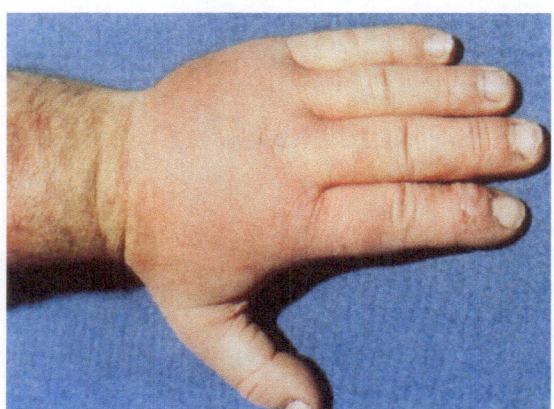

Fig. 4.17. Irreversible chronic traumatic scleroedema and granulomas from sea urchins in a subaqua diver. Reproduced with permission from [6]

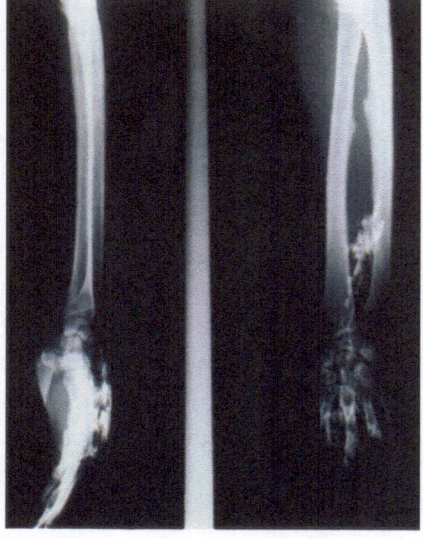

Fig. 4.18. The same case as in Fig. 4.17. Lymphography shows irregular flow and distribution of the contrast medium on the back of the hand. Reproduced with permission from [6]

Table 4.2. Differential diagnosis between spontaneous professional scleroedema (SPS) of the hands and non spontaneous Secrétan's syndrome (NSSS)

	SPS	NSSS
Acrocyanosis	++--	++--
Lesions		
Monolateral	+---	++++
Bilateral	++++	+-----
In association with		
Artefact dermatoses	----	++--
Sea urchin granulomas	++--	----
Age and sex		
Young women	----	++--
Young or elderly men	++++	++--
Worse in winter	++++	----
Lesions persist after abandonment of working activity	++--	----
Psychiatric problems	----	++--

This picture of occupational, hard scleroedema closely resembles Secrétan's syndrome, a cutaneous artefact due to various repeated mechanical stimuli (haemostatic ligatures, occlusive bandaging, traumas), self-inflicted for financial gain (generally to obtain a pension) or psychiatric reasons [19].

Spontaneous, chronic professional scleroedema of the hands must thus be differentiated from self-inflicted complaints (Table 4.2) and other acute or chronic lymphoedemas, such as lymphatic aplasia, recurrent erysipelas, deep thrombophlebitis, angioedema, chilblains, cold urticaria, filariasis, venous obstruction, complications of surgical operations and radiotherapy for breast cancer or other tumours.

Dermatitis from starfish

Starfish (of which there are about 2,000 species) have spines made of calcium carbonate crystals mixed with organic substances (Fig. 4.19). The spines are held erect by special muscular structures. The calcite envelops a glandular tissue that can secrete a toxin: a saponin analogous to that of Holothuroidea with a haemolytic, antibiotic action consisting of steroid glucosides, among which aglycon (in Holothuroidea this is a derivative of lanosterol), of steroid nature, is a derivative of cyclopentanoperhydrophenanthrene [1]. These saponins are strong tensioactive agents that can induce irreversible blockage of neuromuscular transmission. Starfish

Fig. 4.19. *Ofiotrix fragilis* on a sponge

Fig. 4.20. *Echinaster sepositus*

toxin spreads inside the water so that when many such animals are present, contact with the surrounding water can induce a papulo-urticarial itchy eruption. One of the most common starfish in the Mediterranean is *Echinaster sepositus* (Fig. 4.20), which is bright red and particularly common in the Gulf of Naples. The dermatitis can be treated with a lotion with a 0.5% calamine and menthol base.

Some starfish like *Acanthaster planci* (or "crown of thorns"), can inflict a painful sting which may result in granulomatous lesions. This starfish, that lives in the Indo-Pacific region, from Polynesia to the Red Sea, can be as long as 60 cm. *Acanthaster* has from 13 to 16 arms or rays. The outer surface of the body is entirely covered with a series of large, sharp calcareous spines that are very difficult to remove after penetration in the skin. The cutaneous glands also secrete a venom that can induce a severe inflammatory reaction as well as erythema, oedema, vomiting, torpor and sometimes even paralysis.

Another, similar starfish, *Acanthaster elissi*, is present in the eastern Pacific region [20].

The spines of *Acanthaster* easily penetrate gloves and thin shoe soles. They must be surgically removed to avoid the formation of granulomas. The affected area must be immersed in water as hot as is bearable (45° C) for 30-90 minutes or until the pain goes off. Infiltration of the wound with 1-2% lidocaine may relieve the pain. The inflammation can be treated with topical steroids. To prevent damage, the animal should only be handled wearing thick gloves, and only the soft underside of the starfish can be touched with bare hands [20].

Dermatitis from sea cucumbers

The visceral liquid excreted by the sea cucumbers *Cucumaria* and *Stichopus* can irritate the skin and eyes [2].

Some sea cucumbers eat the nematocysts of Coelenterates and these can remain intact for use in personal defence. The treatment of skin reactions due to contact with sea cucumbers should bear this risk in mind and the same means should be adopted as when treating complaints caused directly by Coelenterates.

References

1. Ghiretti F, Cariello L (1984) Gli animali marini velenosi e le loro tossine. Piccin, Padova, 114
2. Fisher AA (1978) Atlas of aquatic dermatology. Grune and Stratton, New York, 27
3. Kaplan EH (1982) Coral reefs. Petherson Field Guides. Houghton Mifflin Company, Boston, 169
4. Banister K, Campbell A (1993) The encyclopedia of aquatic life. Facts on File, New York, 274
5. Angelini G, Bonamonte D (1997) Dermatoses aquatiques méditerranéennes. Nouv Dermatol 16:280
6. Angelini G, Vena GA (1997) Dermatologia professionale e ambientale. Vol I. ISED, Brescia
7. Afa G, Olivetti G (1986) Dermatologia acquatica mediterranea. La Lettera del Dermatologo 5:4
8. Cassano N, Bonamonte D, Angelini G et al (2000) Dermatiti da echinodermi. In: Veraldi S, Caputo R (eds) Dermatologia di importazione. Poletto Editore, Milano, 316
9. Rocha G, Fraga S (1962) Sea urchin granuloma of the skin. Arch Dermatol 85:406
10. Kinmont PDC (1965) Sea urchin sarcoidal granuloma. Br J Dermatol 77:335
11. Meneghini CL (1972) Cases of sea urchin granuloma with positive intradermal test to spine extracts. Contact Dermatitis Newsletter 12:316
12. Burnett JW, Calton GJ, Morgan RJ (1986) Venomous sea urchins. Cutis 38:151
13. Haneke E, Tosti A, Piraccini BM (1996) Sea urchin granuloma of the nail apparatus: report of 2 cases. Dermatology 192:140

14. Baran R, Perrin C (1992) "Shot-gun-like" eruption due to sea-urchin granuloma. Eur J Dermatol 2:506
15. Sala F, Mansi M, Perotta E et al (1988) Granuloma da riccio di mare. Incontri Dermatologici 2:3
16. Vena GA, Foti C, Angelini G (1989) Sindrome di Secrétan in pescatori. In: Ayala F, Balato N (eds) Dermatologia in poster. Cilag S.p.A., Napoli
17. Angelini G, Vena GA, Meneghini CL (1990) Occupational traumatic lymphedema of the hands. Dermatol Clin 8:205
18. Angelini G, Vena GA, Filotico R et al (1990) Linfedema traumatico occupazionale delle mani. Boll Dermatol Allerg Profes 5:75
19. Angelini G, Meneghini CL, Vena GA (1982) Secrétan's syndrome: an artefact oedema of the hand. Contact Dermatitis 8:345
20. Halstead BW (1992) Dangerous aquatic animals of the world: a color atlas. The Darwin Press Inc, Princeton, 45

5 Dermatitis caused by Molluscs

Fig. 5.1. *Thuridilla hopei*

The phylum of Molluscs (from the Latin *mollusca* = a variety of nut with a soft shell) includes about 45,000 species with a wealth of disparate shapes and functions, living in different habitats. Molluscs are present in waters all over the globe; they can be static or mobile, nude (Fig. 5.1) or covered with a protective shell, herbivores or carnivores, microphages or macrophages. The biotoxins isolated from Molluscs up to now have various chemical and pharmacological structures; some are only urticant or have a repellent smell or taste, others are highly toxic and paralyse their prey. In some Molluscs, the filtrating bivalves, the toxins are exogenous and come from phytoplankton. Of the five classes belonging to this phylum, three have the greatest toxicity: Lamellibranchia, Gasteropodia and Cephalopodia (Table 5.1) [1, 2].

There are 11,000 species of Lamellibranchia, which have a bivalvular shell closed by two adductor muscles (the bivalves include mussels and clams, oysters, scallops) and live on the rocks or burrowed in the sand and mud. They are microphages and feed on organic material suspended in the

Table 5.1. The phylum of Molluscs. The toxic species are indicated

1. Class: Lamellibranchs (bivalves)	3. Class: Cephalopods (octopus, squid, cuttlefish)

Species:
  Mytilus galloprovincialis
  Ostrea spp.
  Anomia spp.
	Species:
	  Octopus vulgaris
	  Octopus macropus
	  Eledone moschata
	  Eledone aldrovandi
	  Sepia officinalis
	  Hapalochlaena maculosa

2. Class: Gasteropods (shells)	

A. Subclass: Opisthobranchs
B. Subclass: Prosobranchs
Species: Muricidae (jagged shells)
Species: Conidae (cone shells)
  Conus aulicus
  Conus geographus
  Conus gloria maris
  Conus marmoreus
  Conus striatus
  Conus textile
  Conus tulipa	

water; this material is captured and filtered by the ciliary branchial movements. When attacked by a marine predator, or exposed by low tide to land predators, Molluscs contract the strong adductor muscles and tighten the two valves of the shell, entering an anaerobic state. For this reason they do not produce biotoxins for offence and defence but as they are filtrating microphagic animals (a large mollusc can filter up to 38 litres of seawater per day), they can ingest micro-organisms that produce biotoxins, and thus become poisonous. Obviously, the toxins are innocuous for the molluscs themselves, for unknown reasons. Toxic micro-organisms can be ingested during the so-called "red tides" caused by toxic seaweed species, and the cases of food poisoning after eating contaminated shellfish (clams, oysters) can be attributed to these causes. In some bivalves, saxitoxin has been isolated, a toxin produced by Dinoflagellates whose explosive proliferation is responsible for the red tides.

Gasteropoda may be nude or covered by a shell, generally spiral-shaped. As they are both herbivores and carnivores, they produce a large variety of biotoxins for offence and defence. The best known Gasteropoda belong to the Opisthobranch (*opistho* = posterior) and Prosobranch subclasses: the former have entirely or partially lost their shell while the latter have a highly variable range of hard shells that are generally very beautiful. The best studied species of Prosobranchs are the Muricides (from the Latin *murex* = hard shell) that have lovely jagged shells with many sharp edges and the Conides, hundreds of species with pretty, smooth conical shells of various colours.

An interesting historical note is that Muricides contain a chromogen that oxidizes in contact with the air producing a reddish-purple pigment (purple is 6,6'-dibromoindigo), from which the pigment "Tyrian or Byzantine purple" was extracted as a valuable dye in ancient times. In fact, in various Mediterranean centres where they produced purple, large deposits of *Murex* shells have been found. The toxins produced by the Conides (cone shells) are highly virulent, even if they are little known, and have different pharmacological actions according to the species.

Cephalopod Molluscs (octopus, squid, cuttlefish) immobilize their prey with toxic secretions from their salivary glands. One of the best known members of this class is the octopus, due both to fantastic tales (attacks on scuba divers or ships) and to the fact that in 1950, enteramine (5-hydroxytryptamine or serotonin) was extracted from *Octopus vulgaris* (the most common cephalopod in the Mediterranean). This substance was discovered simultaneously by Vittorio Erspamer in the salivary glands of *Octopus vulgaris* [3] and by Rapoport in beef blood serum (hence serotonin). In addition to serotonin, Cephalopods contain other amines in their salivary glands (tyramine, metatyramine, dopamine, octopamine, histamine) and cephalotoxin, a glycoprotein with a paralysing action on crustaceans [1]. Cephalotoxin has also been isolated in the salivary glands of two Mediterranean octopus species, *Octopus vulgaris* and *Octopus macropus.*

Reactions to Cephalopods

Cephalopods are widespread in all tropical and temperate seas and only in the Arctic and Antarctic are they restricted to a few species. They live in higher density water than normal sea water, generally at depths of less than 100 fathoms (equal to 6 foot or about 1.83 m). The adults of the species prefer rocky bottoms, although they can sometimes be seen in sandy areas. Cephalopods tend to be solitary, pugnacious creatures and when they become aware of the presence of an enemy, they shoot rapidly backwards spewing out powerful jets of water from the siphon at their anterior opening, with their tentacles stretched out horizontally and their head thrust forward. When they are disturbed, they discharge a dark cloud of "ink" that masks their retreat. Sometimes, octopuses may eat their own tentacles. They can reach a maximum length of 9 m, while giant squid can be as big as 19 m.

In the Mediterranean, *Octopus vulgaris* (Fig. 5.2), *O. macropus* (Fig. 5.3), *Eledone moschata, Eledone aldrovandi* and *Sepia officinalis* (Fig. 5.4) are the most common. They are generally inoffensive and timid creatures but thanks to their hard, bony, parrot-like beaks, these Molluscs can take small bites, leaving small, lacerated star-shaped wounds with oedematous margins, which provoke local burning pain that can affect the whole limb. Such wounds can bleed abundantly. Lesions induced by our octopuses, especially *Octopus vulgaris* and *O. macropus,* are not generally followed by systemic disturbances. A toxin with a proteolytic action, eledoisin or muscatin, was isolated by Erspamer from *Eledone moschata.* This causes vasodilation and hypotension in laboratory animals and in man.

Particularly during the summer months, the coasts of Australia are populated by a small, venomous octopus, 10 cm long (including its tentacles), the *Hapalochlaena maculosa* (or *Octopus maculosus*), whose bite may be fatal even for man. It has characteristic markings featuring two blue rings on a brownish-yellow background, hence its name "blue-ringed octopus". When it is aroused, its background colour goes very dark, while the blue rings get brighter and shine like the brilliant eyes on a peacock's tail. A bite from this tiny octopus is lethal in 25% of cases [4]. The bite may initially be painless and symptoms will only onset after 5-10 minutes. A burning pain radiates outward to the whole limb. Typically, there will be copious and prolonged bleeding at the site of the lesion. Local itching may be associated with an allergic urticarial eruption. In serious cases, death may be brought on by muscular and respiratory paralysis.

Fig. 5.2. *Octopus vulgaris*

Fig. 5.3. *Octopus macropus*

Fig. 5.4. *Sepia officinalis*

There is no effective antidote. Treatment is symptomatic and in less serious cases there will be some improvement after 4-10 hours. Complete recovery may take 2-4 days.

Reactions to other Molluscs

Some bivalves (*Mytilus galloprovincialis, Ostrea spp., Anomia spp.*) living in shallower waters can induce cutting wounds of a superficial or deeper nature according to the structure and arrangement of these Molluscs on the rocks.

The Conidae family includes about 400 species, all of the *Conus* genus, and virtually all confined to tropical and subtropical waters. The species *Conus aulicus, C. geographus, C. gloria maris, C. marmoreus, C. striatus, C. textile* and *C. tulipa* have a well-developed venom apparatus and can kill a man. Depending on their size and the nature of their venom apparatus, many other species are potentially harmful to man. Cone shells live in tidal areas from shallow waters down to several hundred metres in depth in different microhabitats: seaweed, coral reefs, sandy bottoms. The most dangerous species to man are the sand dwellers.

A cone shell sting immediately elicits intense burning sensations, torpor and tingling or numbness, which rapidly spread from the affected area to the whole body and are especially pronounced at the lips and mouth. These symptoms are followed by localized ischaemia and cyanosis in the affected area. In severe cases muscular paralysis and coma may onset, although respiratory distress is not usually a feature, and death will ensue due to heart failure within 6 hours of the bite. If the patient survives, the systemic symptoms should subside after about 24 hours but the skin reaction will persist for several weeks.

There is no specific treatment for cone shell stings. Making an incision to suck the poison out of the wound does not seem to help and should be avoided. A good first aid approach is to contain the spread of the venom by a pressure-immobilization technique [2]. When the sting is in a site where the manoeuvre is possible, a gauze or fabric pad about 6-8 cm in diameter and 2-3 cm thick should be pressed against the wound, and bound tightly enough to prevent venous return without hindering arterial flow. The pad must be removed after the victim arrives in hospital, and symptomatic systemic treatment begun.

Because of the great beauty of their shells, Conidae are highly prized by collectors, who sometimes fail to take adequate precautions when handling them. As a preventive measure, gloves must be worn and the shells should be picked up only by their wide, posterior extremity and dropped immediately if the animal extends its radular shaft. They should be handled as little as possible and should not therefore be cleaned of detritus while still alive. Cone shells should never be kept in a pocket as they can inflict stings even through clothing.

References

1. Ghiretti F, Cariello L (1984) Gli animali marini velenosi e le loro tossine. Piccin, Padova, 73
2. Halstead BW (1992) Dangerous aquatic animals of the world: a color atlas. The Darwin Press Inc, Princeton, 41
3. Erspamer V, Ghiretti F (1951) The action of enteramine on the heart of molluscs. J Physiol 115:470
4. Rosco D (1976) Treatment of venomous and poisonous marine animal injuries. Int Soc Aquatic Med Newsletter 2:2

6 Lesions caused by Arthropods

Fig. 6.1. *Palinurus elephas* (lobster)

This work is concerned with marine animals that can induce dermatitis through various pathogenic mechanisms. On this basis, there would be no point in considering those species that lack toxins and can only produce traumatic lesions. However, the need to follow some zoological classification of marine species dictates at least a brief mention of various species lacking in biotoxins, such as the Arthropods.

Unlike the huge number of venomous land species (scorpions, spiders, chilopods, insects), marine Arthropods do not secrete toxins. Of the vast phylum of Arthropods, only most Crustaceans and 5 surviving species of *Xiphosuri* are marine animals.

Crustaceans (crabs, shrimps, lobsters, barnacles) (Figs. 6.1, 6.2), belonging to the Arthropod class, include freshwater forms, and have branchial or integumentary respiration systems. In the larval stages and sometimes even as adults, they account for an important part of plankton and serve as food for many pelagic animals. They have two pairs of antennae and a variable number of articulated, typically cleft appendages. They have a chitinous, often calcified exoskeleton, subdivided into mobile, jointed segments. These are generally fused to form the head, thorax (or cephalothorax) and abdomen and all except the last are supplied with two articulated appendages [1, 2].

Fig. 6.2. *Gnathophyllum elegans* (shrimp)

Apart from a few cases of hermaphroditism, there are separate sexes; parthenogenesis is frequent. Development generally occurs by means of metamorphosis, often through long and complex processes: the nauplius is the typical larval form but a more advanced developmental stage, the zoea, can also hatch from an egg.

These animals, such as crabs and lobsters for instance, appear to be well protected, whatever their size. They have a solid carapace and are armed with two strong claws that end in pincers. Thanks to these mechanical devices, Crustaceans do not need chemical weapons to capture their prey or defend themselves from predators. However, these defences are not always enough. It is well known that an octopus, although it is a mollusc with a soft, bare body, has no difficulty in devouring even large crabs despite the fact that they could amputate undefended tentacles with a single strike of their claws.

Crustaceans, or shellfish, have always been very popular as food. Nevertheless, in some parts of the world (Japan, New Zealand, the Pacific archipelago) various cases of shellfish poisoning have been reported. Studies of the suspected species have shown that in some periods of the year and only in some regions, four of them contain saxitoxin, the toxin that is extracted and purified from bivalve molluscs.

In some species the toxicity has been demonstrated to be due to the presence of *Gonyaulax*, the same Dinoflagellate that is implicated in the red tides that make bivalves poisonous [3]. It is a strange and so far inexplicable fact that the biotoxin should be present only in a few species of Crustaceans.

Reactions to Crustaceans

The class of Crustaceans includes some species that are well known and widespread in the Mediterranean. Any harmful effects are purely mechanical (lacerating wounds) provoked by the claws of large crabs, for instance, such as *Eriphia verrucosa* and *Homarus gammarus*.

Dermatitis of the hands has also been reported in lobster catchers. This is characterized by a pruriginous eruption, complicated by hypercheratosis and ragade-like cracks. There are various underlying pathogenic mechanisms: trauma during manoeuvres for catching and deweeding the crustacean, the contact with sea water and the sensitising action of some seasonal seaweeds.

Mechanical lesions can also be caused by Balani (*balanos* = an acorn), of the Cirripedus crustacean genus (from the Latin *cirrus* = a bifidus appendix) of the barnacle family. These species are non parasitic and have six

Fig. 6.3. Barnacles on a clam

appendages for conveying small planktonic animals to the mouth for food. They are sessile animals covered with a calcareous shell with sharp edges, that can cause cutting wounds of variable severity. Barnacles colonize hard substrates (quays, buoys, the hulls of ships, harbour bottoms) in particularly polluted waters (Fig. 6.3).

References

1. Ghiretti F, Cariello L (1984) Gli animali velenosi e le loro tossine. Piccin, Padova, 111
2. Telford M (1982) Shrimps, lobsters, and crabs. In: Kaplan EH (ed) Coral reefs. Peterson Field Guides. Houghton Mifflin Company, Boston, 150
3. Hashimoto Y, Konosu S (1978) Venoms of Crustacea and Merostomata. In: Bettini S (ed) Arthropod venoms. Springer, Heidelberg New York, Berlin, 13

7 Dermatitis caused by sponges

Fig. 7.1. *Euspongia officinalis*

Sponges, members of the Porifera phylum, normally lie stationary attached to the sea bottom or sometimes the lake bottom. There are 5000 known species of sponges, which vary greatly as to shape, size and colour, ranging from practically invisible to 2 m in length and from pastel tones to bright hues: red, yellow, orange and blue [1, 2].

Sponges are the most primitive pluricellular organisms, composed largely of epithelioid cells; the individual cells, known as choanocytes, are flagellated, have considerable functional autonomy and do not form organs: they have no structure that could be compared to Metazoan structures. They generally feature a sac enclosing a cavity, named the spongocele. The external wall is studded with pores for the penetration of water and the nutrients: bacteria, unicellular seaweeds and organic particles. Sponges are thus filtering, microphagic animals with an endocellular digestive system.

The walls of sponges also consist of calcareous and siliceous spikes of various shapes and of a lattice-like skeleton made of spongin, a scleroprotein. The bath sponges we use are nothing other than the skeletons of Porifera.

Human use of sponges dates back to ancient Greek and Roman times, when those with a high content of spongin were adopted for washing and to pad helmets and armour, while harder ones were employed for abrasive purposes to clean parchment and various objects. Some sponges were used for therapeutic purposes: the ash of *Euspongia officinalis* (Fig. 7.1) was employed to treat goitre, owing to its high iodine content ranging from 2% to 16% of its weight. Natural sponges have now been virtually eliminated in favour of artificial ones.

Sponges have long been known not to be entirely innocuous. The siliceous sponges of the Mediterranean, for instance, (*Suberites domuncula* is one of the most common) contain suberitin, a protein biotoxin with a neurotoxic and haemolytic action [3, 4]. Another highly toxic sponge for fish (which do not feed on sponges except in exceptional cases, and indeed go out of their way to avoid them) is *Latruncula magnifica*, a beautiful red sponge that lives at depths of 6-30 m in the Red Sea. Two toxins, latrunculin

A and B have been isolated from this species [5]; they are macrolide complexes with 16 and 14 atoms of carbon, respectively, in a closed-ring structure and containing a thiazolidone residue.

Owing to their canal system and alveolar structure, sponges offer an ideal habitat for micro-organisms. However, although they live on the sea bed, they are never covered in encrustations and so present a clean, smooth surface (unlike an amphora lying at the bottom of the sea, for instance). This has suggested that sponges may exert some kind of antimicrobial activity and in fact, antibacterial substances such as compounds of phenol bromurates, pyrrole and indole, as well as derivatives of furan, terpenes, sesquiterpenes and diterpenes, have been isolated from M*icrociona prolifera* (a red sponge found on the Atlantic coast).

Sponges not only populate a vast geographical habitat but also very different vertical habitats, ranging from tidal zones to depths of over 2800 m.

Reactions to sponges

Sponge fishing is carried out today in the Mediterranean, in Florida, Cuba and the Bahamas, and dermatitis from sponges has been observed in all of these areas. When detaching them from the sea bed or breaking off their ramifications, the fisherman notices an itching or stinging sensation; within a few hours, pain, oedema and rigidity of the hands onset, together with erythema and blisters. The dermatitis regresses after about 2 days. An eruption of erythema multiforme type may also appear.

The sponges that are harmful to man are notably *Tedania ignis*, *Fibula nolitangere* and *Microciona prolifera* [6, 7].

Tedania ignis (the "fire sponge") is abundant in the Miami area and around the Keys in Florida. Although it has no commercial value, it is a beautiful sponge, both for the way it grows and for its reddish-orange or brilliant vermilion shade, that sometimes verges on orange or orangey-yellow. Contact with this sponge induces itching or stinging and after a few hours, pain, erythemato-bullous lesions, oedema and immobility of the fingers. These symptoms resolve in about 2 days. This sponge may also give rise to an erythema multiforme-like eruption.

Fibula nolitangere (the "poison-bun" or "touch-me-not" sponge) owes its name to the fact that its sting provokes a much more violent reaction than the one described above. It lives in slightly deeper waters and is relatively difficult to recognize, as it resembles many other more common sponges as to size, shape and colour. This sponge generally grows in small clusters and its osculi (pores) are as wide as a man's finger; it is brown on the surface and has a consistency like that of soft bread. *Fibula nolitangere* colonizes the waters of Central America and Australia.

Contact with *Microciona prolifera*, the red sponge present along the east coasts of the United States, causes erythema and oedema. Later, blisters develop on the affected area and have a purulent evolution. Unless it is adequately treated, the dermatitis can persist for several months. Patch testing with a fragment of the sponge will confirm the diagnosis.

Apart from the above contact dermatitis forms due to chemical agents, some sponges can induce traumatic dermatitis as a result of contact with the spicules, which are composed of silicium or calcium carbonate. These spicules penetrate the skin and are difficult to remove. It can be helpful to apply first of all an adhesive plaster to the affected part, to which the spines will attach, and then after peeling off the plaster to apply isopropyl alcohol.

Preventive measures include wearing gloves when fishing for sponges, while scuba divers in tropical waters should wear all-over covering. Treatment of the dermatitis is by applying acetic acid (vinegar) compresses for 15-30 minutes, 3-4 times a day. Isopropyl alcohol (40%-70%) may also be useful. Topical corticosteroids may be applied later. In the presence of marked exudation of the lesions, it will be necessary to administer systemic corticosteroids. The itching sensation can be controlled with systemic antihistamine treatment.

References

1.   Ghiretti F, Cariello L (1984) Gli animali velenosi e le loro tossine. Piccin, Padova, 35
2.   Kaplan EH (1982) Coral reefs. Peterson Field Guides. Houghton Mifflin Company, Boston, 121
3.   Richet C (1906) De l'action toxique de la subéritine (extrait aqueux de *Suberites domuncula*). C R Acad Sci III 61:598
4.   Cariello L, Salvato B, Jori G (1980) Partial characterization of suberitine, the neurotoxic protein purified from *Suberites domuncula*. Comp Biochem Physiol 67:337
5.   Neeman I, Fishelson L, Kashman Y (1975) Isolation of a new toxin from sponge *Latruncula magnifica* in the Gulf of Aquata (Red Sea). Marine Biol 30:293
6.   Fisher AA (1978) Atlas of aquatic dermatology. Grune and Stratton, New York, 45
7.   Halstead BW (1992) Dangerous aquatic animals of the world: a color atlas. The Darwin Press Inc, Princeton, 29

8 Dermatitis caused by algae and Bryozoans

Fig. 8.1. *Caulerpa taxifolia* (green alga)

Dermatitis from algae

About 30,000 different species of marine and fresh water algae micro-organisms (Prokaryotes and Eukaryotes) have been identified, that are classified among vegetable species because they are autotrophic (Fig. 8.1) (Table 8.1) [1]. Many algae, most of which contain chlorophyll, have flagella that enable them to move through the water. They feature highly variable shapes, colours and sizes, ranging from microscopic (just 1 μ in diameter) to gigantic forms reaching 300 foot long (90 m).

As members of the most ubiquitous of the vegetable kingdoms, algae can be found in all environments: geysers, salt water, fresh water, snow, ice, the Arctic Circle. Some are saprophytic or symbiotic, and develop on other plants or animals. Curiously, algae have been found on sea beds down to 12,000 feet (3660 m), although the sunrays can only penetrate ocean waters up to 900 feet (275 m).

Marine and fresh water algae that produce biotoxins and are therefore of medical interest belong to the Cyanophyceae (Cyanobacteria or blue

Table 8.1. Classification of algae. Reproduced with permission from [1]

Prokaryotes	
Bacteriophyta	
Cyanophyta (Cyanobacteria or blue algae)	Cyanophyceae
Eukaryotes	
Rhodophyta (red algae)	Rhodophyceae
Chlorophyta (green alghe)	Chlorophyceae Prasinophyceae Charophyceae
Euglenophyta	Euglenophyceae
Xanthophyta (yellow algae)	Xanthophyceae
Bacillariophyta (diatomea)	Bacillariophyceae
Chrysophyta (yellow-brown algae)	Chrysophyceae
Phaeophyta (brown algae)	Phaeophyceae
Pyrrhophyta (Dinoflagellates)	Dinophyceae
Cryptophyta (Cryptomonads)	Cryptophyceae

algae) (Prokaryotes) and Dinophyceae (Eukaryotes) classes. However, even among these two classes there are only a few toxic species. Cyanobacteria, that are rounded or filament-like in shape, live in fresh or salt waters. In some periods of the year they reproduce in great masses, giving rise to the so-called "blooms" in lakes and the seas. There are few fresh water species of Cyanophyceae, notably *Microcystis aeruginosa*, *Anabaena flos-aquae* and *Aphanizomenon flos-aquae*. The best known toxic marine Cyanophyceae species include *Lyngbya majuscola*, *Oscillatoria nigroviridis*, *Spirulina platensis* and *Schirotrix calcicola*. Some of these live in coral reefs and are responsible for the sudden outbreaks of poisonous fish in the area.

The so-called "red tides" (an inappropriate term as they have no connection with the tide) consist of the sudden appearance of turbid, reddish zones on the sea surface. This is due to the immense number of micro-organisms present and the colour of the pigment in their cells. Blue and green algae, diatomea and protozoa cause the red tides: the vast number of toxic Dinoflagellates causes mass death of the fish and other species present in the zone. The red tides are not the same as the blooms, as the former are unforeseeable, whereas the latter appear regularly once or twice a year. They are the result of a higher content of nutrient substances in the water, which in turn causes proliferation of the plankton populations. In the red tides, the water becomes a concentrated suspension of micro-organisms, that virtually always belong to a single species, unlike the variety present in the blooms. Red tides disappear as fast as they appear: they regress in a couple of days, although their effects last a long time. The most common Dinoflagellate species are *Gonyaulax catenella* (containing the toxin gonyautoxin), *G. tamarensis*, *Gymnodinium breve* (containing brevetoxin). The species that cause fish to become poisonous in tropical seas are *Gambierdiscus toxicus* and *Prorocentrum lima*. Some edible bivalves are not killed by the toxins but become carriers, thus posing a serious health problem for man [1].

Swimmer's itch

A Cyanophycea alga that causes a dermatological condition is *Lyngbya majuscola*, present in rivers, lakes, pools and oceans. It belongs to the Oscillatoriacea family and grows as greenish-blue, very long, slender filaments (looking like hairs). It covers sandy sea bottoms or lives attached to other algae. Not all races are toxic, and those in some zones may be toxic but those in zones a few miles away innocuous. *Lyngbya* is abundant in tidal areas and up to a depth of 100 feet (30 m), and is therefore a danger to all bathers. Two toxins that cause painful dermatitis have been isolated and purified from this species: lyngbyatoxin A and debromoaplisiatoxin [1].

In the summer of 1958, *Lyngbya* caused a dermatitis epidemic (125 cases) among Hawaiians living in one coastal area of the island of Oahu (capital, Honolulu) [1-3]. Only a few minutes after bathing in the turbid algae-filled water, intense itching and burning sensations onset, followed after 3-8 hours by blisters leaving painful erosions especially in the perineal region (the genitals and ano-perianal area). The eruption affects the areas covered by the swimming costume; in males it is especially severe on the scrotum and in females on the breast. Some cases of skin reactions to freshwater blue-green algae have also been reported [4, 5]. This dermatitis must be differentiated from "seabather's eruption" and is treated with systemic corticosteroids. Suitable prophylaxis is for swimmers to remove their wet costumes as soon as they come out of the sea and shower using soap.

Dermatitis from other seaweed species

Some French authors reported observation in a fisherman of a simultaneous contact allergy to a bryozoan (*Electra pilosa*) and to *Sargassum muticum*, a brown seaweed (Phaeophyceae) that originates on the Japanese coasts of the Pacific [6]. This seaweed spread to Europe about 15 years ago, possibly as a result of importing Japanese oysters. *Sargassum muticum* colonies have an explosive growth from March to October, the maximum fertility period being in June and July. Apart from competing with the local marine species, they can damage small boats, owing to their great size. At present, *Sargassum muticum* is the only erect species on the European coasts that can reach a height of 10 m. The reason why it can proliferate so widely is because it attaches to various supports and has a remarkable regenerative power (new growth can start from a small fragment that has broken off). This seaweed can be considered toxic as it secretes phenol compounds.

Dermatitis from mucilaginous aggregates

During the summer months from 1989 to 1992, large quantities of mucilaginous aggregates of algal origin appeared in vast zones of the Adriatic Sea. The marine organisms that can secrete mucilage are the Dinoflagellates and above all the diatomea. As well as damaging the marine ecosystem, these aggregates can cause food poisoning in man through intoxication of the fish. Just swimming should not be harmful, although mucilaginous aggregates attract pathogenic bacteria that are then incorporated in the gelatine [7].

Kokelj and coll. [7] reported the case of a woman who developed intensely itchy erythemato-vesiculo-pustular lesions of the limbs a few hours after swimming in the Gulf of Trieste in August 1991. Histological examination of one of the lesions showed spongiosis, vasodilation and neutrophilic infiltration of the superficial derma. The dermatitis regressed after 6 days of topical corticosteroid treatment. Patch tests with samples of water containing the mucilage aggregates gave negative results in 10 healthy volunteers. The authors believe that this rare observation should be included among irritant contact reactions.

Protothecosis

Skin protothecosis is an exceptional infection induced by seaweed, that mainly affects subjects with an impaired immune system and features various clinical pictures. Most cases are caused by achlorotic algae of the *Prototheca* species [8-12]. These are unicellular, colourless (because they lack chlorophyll) micro-organisms classified as an achlorotic mutant of green seaweed of the *Chlorella* genus. *Prototheca* is a eukaryotic non micelial organism, spherical or round, varying from 2 to 25 μ in size, that reproduces asexually by internal cleavage (endosporulation), forming morula with up to 20 endospores. It was isolated in 1894 and classified in 1916 [13].

Prototheca's natural habitat is tree mucilage and sewage; it can also be found in the soil, lakes, ponds, in cats and dogs and in cow's milk. The organism has also been isolated from human nails, skin, expectorate and faeces without any infection being present [14].

At present, 3 species of *Prototheca* are known: *P. stagnora*, *P. wickerhamii* and *P. zopfii*. *Prototheca moriformis*, *P. portoricensis*, *P. ciferii*, *P. ubrizsyi* and *P. segbwema* are identical to *P. zopfii*. In most cases, the causal infectious agent is *P. wickerhamii*, while there have been few reports of infection from *P. zopfii* [15, 16].

About 80 cases of protothecosis have been reported in the literature in various parts of the world: Europe, Asia (Japan, China, Thailand, Vietnam), Africa, Panama, Oceania and the southeast of the United States. However, the infection seems to be most common in tropical areas [17, 18].

Protothecosis most often manifests as lesions confined to the cutaneous or subcutaneous layer; in about 25% of cases it presents as olecranon bursitis. Systemic forms are very rare, and exclusively affect subjects with impaired immune response [18].

The infection is transmitted mainly as a result of trauma. However, the pathogenic potential of these algae seems to be relatively low, and infection seems to be induced by a state of local or systemic immune depression. In fact, apart from a very few exceptions, skin protothecosis is observed in subjects with a debilitating disease or immune deficiency. It has been reported to be associated with the acquired immune deficiency syndrome (AIDS), with immunosuppressive treatment and with alterations in function of the polymorphonuclear neutrophils [19, 20]. Forms of superficial and subcutaneous protothecosis may also be secondary to intralesional corticosteroid therapy [21, 22]. In laboratory animals, the infection has been reproduced only after pre-treatment with hydrocortisone [17]. Unlike cutaneous and systemic forms, olecranon bursitis affects immunocompetent subjects. In about 50% of cases local trauma, such as car accidents, falls or surgical operations, has been demonstrated to play a fundamental role in inducing the infection [16].

The first case of human protothecosis was described by Davies and coll. in 1964 [15]. Of the 45 cases reported by Nelson and coll. [17], 28 (62%) presented cutaneous or subcutaneous infections. Most of these consisted of papular or eczematous lesions confined to the face and limbs. These lesions, single or multiple with a slow evolution, had a centrifugal spread. Other reported types of lesions include vesiculae, cellulitis, verrucous nodules, pustules, ulcerative papules and plaques. It is not known how long the incubation time of the infection is but it is thought to take some weeks or months. As there are no specific clinical pictures, diagnosis is made by searching for the causal agent in the tissues, and confirming this in culture. As the micro-organism reproduces by endosporulation the various stages are easily identified in the tissues. The cellular elements have a hyaline wall that is little stained by haematoxylin-eosin, but responds better to PAS, mucicarmine or Gomori (methenamine-silver). The single cells have a basophilic content demonstrated by haematoxylin-eosin and measure from 2 to 25 µ depending on the species. Cells in the initial sporulation phase show cleavage and are slightly bigger. Mature forms in advanced sporulation are the most characteristic, measuring over 20 µ and resembling morula.

The *Prototheca* species grows within 48 hours in Sabouraud dextrose agar culture medium at between 25° and 37° C in the absence of cyclohexamide, forming smooth, yeast-like, creamy-white colonies. The individual forms can be identified by inoculating them in the culture medium (sucrose, trialose, inusitol and propanolol) over 14 days. More rapid identification and differentiation of the species can be achieved with immunofluorescence tests of cultures and tissues.

Histological examination demonstrates many isolated or aggregated organisms among the collagen fibres, within the histiocytes in the papillary and reticular derma, in the annexes and the epidermis. In haematoxylin-eosin stained sections, the organisms appear strongly basophilic with a pale halo. The epidermis is hyperplastic and an inflammatory infiltrate is evident, scattered or organized as granulomas.

Clinical diagnosis is not easy. A skin trauma can favour entry of the organism. From the aetiological point of view, *Prototheca* must be differentiated from unicellular green algae (being an achlorotic mutant of these), that can in exceptional cases cause human infection [23]. The two organisms have a strictly related size, shape and reproduction method and similar staining characteristics. Green algae cells are differentiated on the basis of the presence of cytoplasmic granules, which are lacking in *Prototheca*. However, these granules cannot be revealed by haematoxylin-eosin staining (which cannot therefore differentiate between the two organisms) but only by PAS or Gomori staining.

Treatment of skin protothecosis is quite difficult, especially in immune-deficient patients. The lesions can persist for years and can also spread. Exceptionally, they may also resolve spontaneously [24]. There is no known elective treatment. Localized lesions can be surgically removed. Oral ketoconazole has been found efficacious in some cases; intravenous amphotericin B, alone or in association with oral tetracycline has had variable success. In a few cases reported recently, itraconazole [25] and fluconazole [21] gave good results.

Dermatitis from Bryozoans

Up to now, 4000 species of Bryozoans (*brion* = moss) have been identified. These belong to the animal kingdom and are very ancient invertebrates, which appeared in the sea in the primary era. For a long time they were mistakenly taken to be algae or corals. They grow in colonies and each colony (zooarium) consists of masses of units, ranging from a few, up to thousands of individual animals. Each animal measures about 1 mm and appears as a tiny cell with a chitinous wall, a visceral portion starting with a buccal orifice surrounded by a crown of tentacles, and its ciliary movements attract food particles suspended in the water.

The colonies live attached to the sea bottom, and assemble as coral-like masses encrusting rocks (Fig. 8.2), shells or seaweed, or free-floating masses with filament-like or tree-like extensions like those of seaweed. Various Bryozoans can cause contact dermatitis [6, 26-28]. In most cases, *Alcyonidium gelatinosum*, a filament-like zooarium, is responsible, and occasionally

Fig. 8.2. *Sertella beaniana* (ribbon alga)

Alcyonidium hirsutum that builds encrusting colonies or *Alcyonidium topsenti*; in exceptional cases *Flustra foliacea*, a free-floating or encrusting zooarium, and *Electra pilosa*, an encrusting zooarium, can be implicated.

Alcyonidium gelatinosum is one of the oldest animals on the planet and looks like a yellow-green-brown alga (hence the name " algue pain d'épice" given to this animal by fishermen from Le Havre). It lives in colonies attached to hard substrates (rocks, shells, gravel, stones) in filaments about 20-30 cm long. Each colony consists of a large number of hermaphrodite individual creatures. These produce larvae that then attach to the same filament or to other substrates creating new colonies. The animal feeds on plankton and proliferates during the summer season, whereas it becomes rarefied during the winter as from October. *Alcyonidium gelatinosum* is widespread in the Northern Hemisphere, above the 45th parallel of north latitude, and especially in the Atlantic, the Baltic, the North Sea, the Arctic and the North Channel. It is also present in the Southern Hemisphere, in the coral zones of the Pacific, in Australian seas and more rarely in the Mediterranean and the Adriatic Sea.

Electra pilosa builds a zooarium of white colonies that encrust various substrates. It is common in most of the seas in the Northern Hemisphere (the North Sea, the Channel, the Atlantic, the Mediterranean) at depths of between 15 and 18 m and it proliferates during the warm seasons [6].

Contact dermatitis from Bryozoans affects fishermen and is quite disabling. It was first observed in the North Sea (hence its first name "Dogger Bank Itch", from the Dogger Bank area in the North Sea, given to the disease by Bonnevie but now no longer used) [26], and then reported in the eastern part of the Channel, in the Bay of the Seine [27] and in Polynesia.

This professional eczema typically affecting fishermen manifests by means of an allergic mechanism of delayed type, that recurs each summer after the initial sensitisation. This can onset even on first exposure. Fishermen come into contact with "sea moss" or "sea mats" when they pull their nets on board the boat and find them jumbled in with the fish; sometimes large quantities are present and are then thrown overboard. The clinical picture of the dermatitis features both dry, fissuring and acute, exudative lesions. The hands and forearms are first affected through direct contact with the Bryozoa; the face and neck may also be involved through airborne contact (airborne allergic contact dermatitis) with drops of sea water containing the allergenic material. With successive exposure, the dermatitis may become generalized and manifest as a blistering, oedematous eruption.

The allergen responsible is 2-hydroxyethyldimethylsulphoxonium in the case of *Alcyonidium gelatinosum* [29]. Epicutaneous tests are made with fragments of live bryozoans just harvested, with sea water containing the allergen and with aqueous and acetonyl extracts of sea moss. Histological tests have shown intraepidermal spongiosis and a perivascular inflammatory dermal infiltrate [27].

References

1. Ghiretti F, Cariello L (1984) Gli animali marini velenosi e le loro tossine. Piccin, Padova, 17
2. Fisher AA (1978) Atlas of aquatic dermatology. Grune and Stratton, New York, 52
3. Grauer FH, Arnold HL (1961) Seaweed dermatitis. Arch Dermatol 84:720
4. Cohen SG, Reif CB (1953) Cutaneous sensitization to blue-green algae. J Allergy 24:452
5. Heise HA (1951) Microcystis: another form of algae producing allergenic reactions. Ann Allergy 9:100
6. Jeanmougin M, Lemarchand-Venencie F, Hóang XD et al (1987) Eczéma professionnel avec photosensibilité par contact de Bryozoaires. Ann Dermatol Venereol 114:353
7. Kokelj F, Trevisan G, Stinco G et al (1994) Skin damage caused by mucilaginous aggregates in the Adriatic sea. Contact Dermatitis 31:257
8. Bonamonte D, Cassano N, Vena GA et al (2000) Prototecosi. In: Veraldi S, Caputo R (eds) Dermatologia di importazione. Poletto Editore, Milano, 134
9. Sudman MS (1974) Prototecosis. Am J Clin Pathol 61:10
10. Mayhall CG, Miller CW, Eisen AZ et al (1976) Cutaneous protothecosis. Arch Dermatol 112:1749
11. Angelini G, Vena GA (1997) Dermatosi da agenti marini. In: Angelini G, Vena GA (eds) Dermatologia professionale e ambientale. Vol I. ISED, Brescia, 202
12. Monopoli A (1995) Cutaneous protothecosis. Int J Dermatol 34:766
13. West GS (1916) Algae. Cambridge University Press 1:475
14. Sonk CE, Koch Y (1971) Vertreter der Gattung *Prototheca* als Schmarotzer auf der Haut Mykosen 14:475
15. Davies RR, Spencer H, Wakelin PO (1964) A case of human protothecosis. Trans R Soc Trop Med Hyg 58:448

16. Mendez CM, Silva-Lizama E, Logemann H (1995) Human cutaneous protothecosis. Int J Dermatol 34:554
17. Nelson AM, Neafie RC, Connor DH (1987) Cutaneous protothecosis and chlorellosis, extraordinary "aquatic-borne" algal infections. Clin Dermatol 14:475
18. Huerre M, Ravisse P, Solomon H et al (1993) Protothécoses humaines et environnement. Bull Soc Pathol Exot 86:484
19. Woolrich A, Koestenblatt E, Don P (1994) Cutaneous protothecosis and AIDS. J Am Acad Dermatol 31:920
20. Wirth FA, Passalacqua JA, Kao G (1999) Disseminated cutaneous protothecosis in an immunocompromised host: a case report and literature review. Cutis 63:185
21. Kim ST, Suh KS, Chae YS et al (1996) Successful treatment with fluconazole of protothecosis developing at the site of an intralesional corticosteroid injection. Br J Dermatol 135:803
22. Walsh SV, Johnson RA, Tahan SR (1998) Protothecosis: an unusual cause of chronic subcutaneous and soft tissue infection. Am J Dermatol 20:379
23. Jones JW, Fadden HW, Chandler FW et al (1983) Green algal infection in a human. Am J Clin Pathol 80:102
24. Dogliotti M, Mars PW, Rabson AR et al (1975) Cutaneous protothecosis. Br J Dermatol 93:473
25. Tang WYM, Lo KK, Lam WY et al (1995) Cutaneous protothecosis: report of a case in Hong Kong. Br J Dermatol 133:479
26. Bonnevie P (1948) Fishermen's "Dogger Bank Itch" allergic contact eczema due to coralline *Alcyonidium hirsutum*, the sea-chervil. Acta Allergol 1:40
27. Audebert C, Lamourex P (1978) Eczéma professionnel du marin pêcheur par contact de Bryozoaires en Baie de Seine. Ann Dermatol Venereol 105:187
28. Jeanmougin M, Janier M, Prigent F et al (1983) Eczéma de contact avec photosensibilité à *Alcyonidium gelatinosum*. Ann Dermatol Venereol 110:725
29 Martin P (1983) Dermatoses due to bryozoans. In: Kukita A, Seiji M (eds) Proceedings of the XVIth International Congress of Dermatology. University of Tokyo Press, Tokyo, 503

9 Dermatitis caused by aquatic worms

Fig. 9.1. *Serpula vermicularis* (Polychaeta)

W orms are marine and land invertebrates with a long, soft, contractile body and no limbs. At present these organisms are classified in the phyla of Platyhelminthes, Nemertea, Nematoda, Annelida and many others (Table 9.1). The marine forms of these phyla have been investigated for toxins, and the chemical nature of some of those found has been identified [1].

The Annelid phylum, consisting of segmented worms, features organisms with an elongated body and pairs of bristles. The external body covering is a thin, non-chitinous cuticle. Some species have well developed chitinous jaws. Annelid worms are cosmopolitan and can live in seawater, freshwater and earth.

The Annelida phylum includes 4 classes of worms [2]:

1. Hirudinea class: leeches.
2. Oligochaeta class: worms with few bristles, like earthworms.
3. Archiannelida class: primitive worms that are rarely observed.
4. Polychaeta class: worms covered in bristles, that are the most common marine forms (Fig. 9.1).

The few venomous Annelid species are metameric worms belonging to the Polychaeta class, and are all marine animals. Nereistoxin, a tertiary

Table 9.1. The phyla of some toxic salt and fresh water worms

1. Phylum: Platyhelminthes	2. Phylum: Annelida

Family: Schistosomatidae (cercariae)
 Species:
   Trichobilharzia ocellata
   Trichobilharzia stagnicolae
   Trichobilharzia physellae
   Gigantobilharzia huronensis
   Schistosomatium douthitti
	A. Class: Hirudinea (leeches)
	 B. Class: Polychaeta
	 Species:
	   Hermodice carunculata (dogworm)
	   Aphrodite aculeata (sea mouse)
	   Nereis diversicolor
	   Lumbriconereis impatiens

	3. Phylum: Nematoda

	Species:
	   Ancylostoma spp.
	   Oncocherca volvulus

amine with a cyclic disulphuric group, has been isolated from *Lumbriconereis heteropoda* [3, 4]. The Nemertea, that are similar to Platyhelminthes but have more complex organs and apparatus, range in size from a few mm to several metres and live in seaweed and among rocks. They are equipped with a long, generally eversible proboscis: this is everted to capture the prey, which is then killed by contact with a slimy liquid that the animal secretes. The substances nemertine, amphiporine, nemertelline and anabasin have been isolated in this liquid. The latter three are pyridinic toxins with a nicotine-like action [5].

Cercarial dermatitis

Cercarial dermatitis ("swimmer's itch" or schistosome dermatitis) is an acute inflammatory non contagious eruption caused by skin penetration of cercariae, of the Schistosomatidae family (Platyhelminthes) [6-9]. The clinical picture, induced by non human-parasitizing schistosomes, is different from visceral and cutaneous schistosomiasis or bilharzias, caused by the human-parasitizing species *Schistosoma mansoni, S. japonicum* and *S. haematobium*.

Cercarial dermatitis is well known everywhere as it is widespread in the East and West, in the Arctic, in tropical and temperate zones and in salt and freshwater. As well as in bathers and underwater divers, it can be observed in subjects working with freshwater for irrigation (agricultural workers and rice growers) [10, 11]. This disease, that is referred to by various synonyms, must be differentiated from "seabather's eruption".

The schistosomes that induce this form of dermatitis are blood parasites of birds and mammals, but in their ecological cycle, man can become a chance host. The cycle starts with hatching of the eggs contained in the faeces of infested animals. Through contact with the ciliated miracidia from the eggs, some varieties of shellfish can become infested and act as intermediate hosts. After a period of incubation in Molluscs, the cercariae are released into the water; through their fork-like extremity these parasites then batten on specific warm-blooded animals used as hosts (sea birds, mice, sparrows), which in turn evacuate the eggs and complete the cycle. Occasionally, man can contract cercarial dermatitis and enter the life cycle of these parasites.

The most common of these are *Trichobilharzia ocellata, T. stagnicolae, T. physellae, Gigantobilharzia huronensis* and *Schistosomatium douthitti* [9]. The Mollusc intermediate hosts most often belong to the *Lymnaea, Physa, Planorbis, Polyplis* and *Stagnicola* species. According to the target host, three forms of cercarial dermatitis can be distinguished, two freshwater forms and a seawater form.

Dermatitis from freshwater cercarias. This form, whose target host is a bird, has also been reported in Italy among female workers in rice fields [11]. This seasonal affliction has been reported in the Padania Plain during periods coinciding with the rice-picking season but technological changes in rice growing methods have made it an exceptional observation. Summer cases among bathers in streams containing cercariae have also been described [11].

Dermatitis from freshwater cercarias. This affliction is due to cercarias whose hosts are buffaloes, sheep and goats and has been reported in workers in oriental rice fields (India, Malaysia, China) and Austral Africa.

Dermatitis from saltwater cercarias. The target host of cercarias causing this dermatitis is a sea bird, while the intermediate hosts are marine Molluscs. This form has been described in bathers in infected waters in the USA, Australia and Hawaii.

Clinical manifestations onset as hypersensitivity phenomena in allergic subjects. The allergic reaction is elicited by the process of destruction of the cercarias, that occurs in the epidermis because cercarias seem to be unable to penetrate beyond the papillary derma.

An initial stinging sensation is followed by the rapid development of wheals, which resolve in about half an hour leaving maculae. Within a few hours these turn into very itchy papules which reach their most severe form of expression by the second or third day. The papules resolve in 1-2 weeks but the dermatitis can be complicated by excoriations due to scratching and secondary infections with the formation of pustules.

Individual prevention is obviously important (wetsuits, rapid drying by energetic rubbing with rough towels), as is environmental clean-up (antiparasite treatments). Mild dermatological complaints are treated with antiseptic compresses and systemic antihistamine therapy. More severe eruptions require systemic corticosteroids and antibiotics.

Reactions to leeches

Leeches are freshwater segmented worms belonging to the Hirudinea class (from the Latin *hirudo* = bloodsucker) and the Annelida phylum. These dark green invertebrates have a slender body about 5-7 cm long, attach to the skin and suck blood until they are gorged and have doubled their volume, whereupon they drop to the ground. Their saliva has anticoagulant, fibrinolytic and local anaesthetic properties, causing the victim to bleed freely without feeling pain.

In the ventral position at the cephalic level, leeches have a terminal sucker which attaches to the victim, and in the dorsal position they have five pairs of punctiform eyes. They have a cutaneous respiratory system and live in ponds and streams, feeding on the blood of vertebrates and especially mammals. There are sea and land leeches (the latter live in tropical rain forests). In infested waters, man can be bitten during the summer months [12].

With its bite, the leech injects an anticoagulant, hirudin, together with other antigenic substances whose nature is still unknown. In non sensitised subjects, the wound bleeds and heals slowly. In allergic subjects, urticarial, blistering or necrotic and even anaphylactic reactions may onset. Contact sensitisation is only exceptionally observed [13]. Leeches can be vectors of various agents that are pathogenic to man, and reside in the bowel [14, 15].

In some Middle Eastern and Oriental countries, leeches are shown off in large glass containers for sale in the market for medicinal purposes. They must not be forcibly detached from the skin as the jaw may remain in the wound; removal from the host is favoured by the application of heat (lighting a match under it), alcohol or strong vinegar.

Modern medicine has reinstated the use of leeches for some particular indications in microsurgery and plastic surgery, to prevent venous congestion and thrombosis of skin flaps [15].

Dermatitis from Polychaetes

Polychaetae annelid worms are metameric, covered in bristles, and have a cylindrical body. Their segments form many somites or body units, each equipped with paddle-like appendages, or parapodia, that are covered with setae. Tentacles extend out from the region of the head. There are two major groups of Polychaetes: Errantia, that are free-moving animals, and Sedentaria, that live in burrows. The toxic species belong to the errant group.

There are estimated to be over 6200 species of these worms, present in all seas, from the tidal zones up to depths of 5000 m. They burrow in the mud, corals, palings or under rocks. Few species live in the open sea. Most of these worms are from 5 to 10 cm long, although different species range from only 2 mm, to the giant Australian kind measuring 1 m or more. Many Polychaetes are extremely beautiful, being iridescent and red, pink or green or a combination of various colours.

The Polychaeta venom apparatus includes bristles and jaws. The members of the *Chloeia*, *Eurythoe* and *Hermodice* species have mordant chitinous bristles projecting out of the parapodia.

Fig. 9.2. *Hermodice carunculata* (dogworm)

When Polychaetes are at rest, the bristles are retracted and appear very short but when they are alarmed, the bristles are rapidly extended and the worm takes on the appearance of a bristly mass. The *Glycera* genus has a long extensible proboscis, equipped with fangs connected to toxin-secreting glands, and can inflict painful bites [16].

Reactions to Polychaetes

The best known toxic species are *Chloeia flava* (present on the coasts of Malaysia), *C. viridis* (the Caribbean), *Eurythoe complanata* (Mexico and the tropical Pacific), *Glycera dibranchiata* (the east coasts of the United States and Canada), *Eunice aphroditois* (tropical seas), *Hermodice carunculata* (the Gulf of Mexico) (Fig. 9.2).

In the Mediterranean, *Hermodice carunculata* (the "dogworm") and *Aphrodite aculeata* (the "sea mouse") are widespread. The former is especially abundant in the Aegean and the seas around Sicily; its habitat is hard, dark substrates (grottoes, rocks, wrecks). It is elongated, of variable length and size, with setae at the sides of each metamere. Penetration of the setae in the skin causes intense, pruriginous and painful erythema and oedema, and loss of sensation of the affected part [17].

When a joint is involved, painful hydrarthrosis may onset, with diminished function. Unless the bristles are immediately removed (using a sticking plaster), they may be expelled through the development of a granulomatous or purulent inflammation.

The purely mechanical effect of the toxins may be associated with a tox-

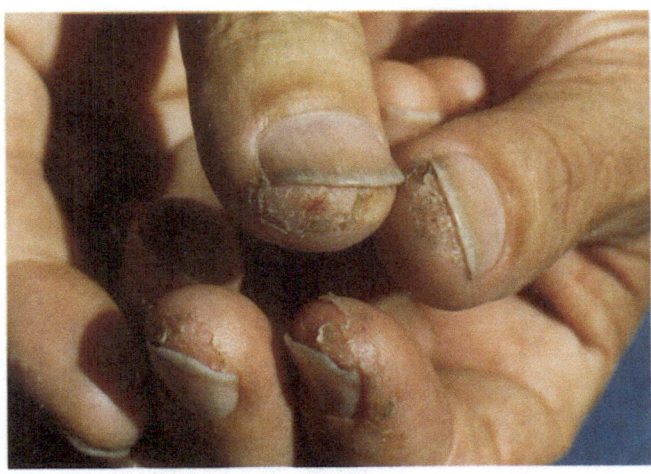

Fig. 9.3. Contact dermatitis to bait. Hypercheratosis and ragades of the fingertips with onycholysis. Reproduced with permission from [18]

ic action mediated by as yet unidentified biotoxins. These induce anaphylactic reactions affecting cardiac and respiratory function, above all, as observed in two professional scuba divers that accidentally came into contact with a dog worm 8 miles to the east of Lampedusa [19].

The bristles of the sea mouse, that is a stout, virtually elliptic shape, have a similar action but sea mice prefer sandy and muddy sea bottoms with a downwards slope.

Contact dermatitis from bait

Fishermen using lines, "Sunday" and leisure time fishing enthusiasts may, albeit rarely, present a peculiar contact dermatitis. We have observed one such case [20]. A thirty-two year old civil servant and line-fishing enthusiast developed contact dermatitis of the left hand and first finger of the right hand, that recurred for two consecutive years, lasting from June to September, the months when he could indulge his hobby most frequently. The affliction involved the fingertips, proximal nail folds and nails and manifested as desquamation, ragades and onycholysis (Fig. 9.3). The patient was in no doubt about the cause of the dermatitis, that was very painful: it was provoked by a sea worm used as bait. Each time, the affliction onset 10-24 hours after contact with this bait and regressed when he stopped using it. He had no problems when using other bait (shrimps).

The incriminated worm was *Nereis diversicolor* (sea Scolopendra), an annelid of the Polychaeta class that is very common in the North Sea, the

Fig. 9.4. *Nereis diversicolor.* Fishing bait on sale. Reproduced with permission from [18]

Fig. 9.5. *Bernardo* hermit crab (paguro) and *Calliactis parasitica* (sea anemone)

Channel and the Mediterranean. The *Nereis* bait came from Normandy and is also sold in Italy in shops selling fishing articles, under the name "saltarello coreano" (it is widely cultivated in Korea) (Fig. 9.4).

Contact dermatitis brought on by bait is very rare and has been observed up to now only on the Mediterranean coasts. Some French authors reported 3 cases induced by *Nereis diversicolor,* by a crustacean of the Paguridae genus (the hermit crab) (Fig. 9.5) and by a lugworm, respectively [21-23]. The first of the 3 cases was dubbed "escavenite" [21].

Other cases have been observed on the Ligurian Riviera [24] and in Spain [25], caused by *Nereis diversicolor* and *Lumbriconereis impatiens.*

While hooking fragments of the bait, the coelomic liquid impregnates the distal portion of the fisherman's fingers causing the above clinical picture (Fig. 9.6). This affliction is commonly considered to be a form of pro-

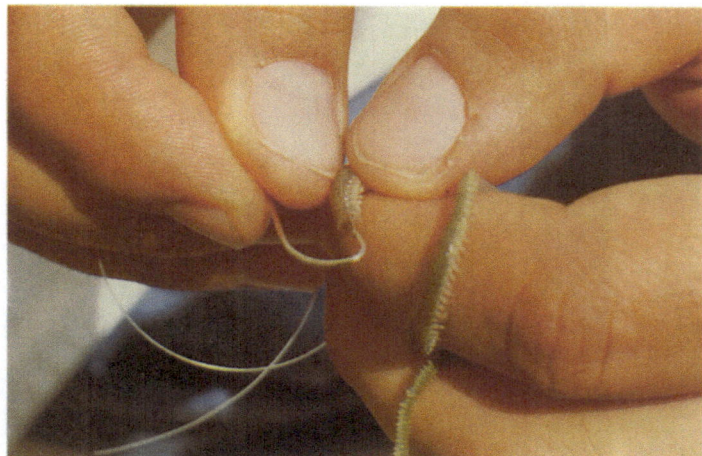

Fig. 9.6. Threading the bait on the hook. The coelomic liquid of the annelid worm segment impregnating the fisherman's fingers induces dermatitis. Reproduced with permission from [18]

tein contact dermatitis, similar to that provoked by some foods or chemical substances. Nereistoxin, a tertiary amine with a cyclic disulphate structure exerting a powerful insecticide action, has been isolated from some species of annelid Polychaetes [4, 5].

Dermatitis from Nematodes

The Nematoda phylum, of non segmented round worms, includes about 12,000 species that are widespread throughout the world. Most of these species live in damp soil and fresh or saltwater, while some are parasites of animals and plants. They vary in size from 0.05 mm to about 1 mm, but are mostly between 0.1 and 0.2 mm.

Larva migrans cutanea

In man, larva migrans cutanea (or "creeping eruption"), a dermatosis induced by the migration of parasites through the skin, is caused by nematode larvae, including the *Ancylostoma brasiliense* species. *Ancylostoma caninum*, the dog parasite, *Uncinaria stenocephala*, the European variant of the same worm, and *Bumostomum phlebotomum*, a bovine parasite, are more rarely implicated. This dermatosis can exceptionally be induced by other worms such as *Anatrichosoma cutaneum*, *Necator americanus*, *Dirofilaria repens*, *Spirometra miathostoma*, *Loa loa*, *Ancylostoma duodenale*, *Strongyloides stercoralis* and *Gnathostoma hispidum* [26-28].

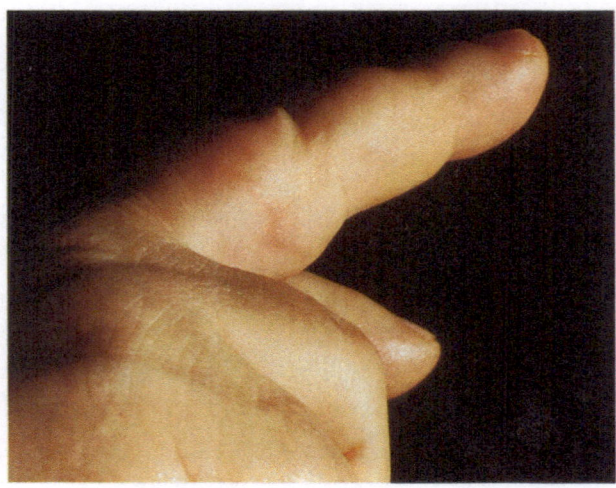

Fig. 9.7. Larva migrans cutanea

These Nematodes are natural parasites of dogs, cats and some wild animals. The affliction is most commonly observed in tropical and subtropical areas, particularly South Africa, the Far East, Central America and southern parts of the United States. In Europe and in temperate zones in general, autochthonous cases of creeping eruption are less frequent [29-33], although in these zones it is possible to observe imported cases in subjects who have stayed in regions where the disease is endemic [34]. Of the 12 cases observed in our department only one was imported, seen in a nun who had been in Kenya (Fig. 9.7).

The parasite lodges in the bowel of the host animal. The fertilized eggs expelled with the animal faeces develop into embryos and thence into the various stages of larva up to the infesting stage (strongyle or III stage larva). The latter enter the animal through the hair follicles or continuous skin solutions and travel in the circulation to the lungs, bronchi and trachea, and to the intestinal tract through swallowing them.

Man can become part of the life cycle as a temporary host. The larvae are unable to penetrate beyond the derma, probably because they lack the necessary enzymes [35]. Hence, in man the larvae migrate within the epidermis or between the derma and the epidermis [36]. In these areas, progression of the larvae varies from a few mm to some cm per day. The larvae migrate for biological purposes, since they need to reach the host intestine where the second phase of the life cycle will terminate and the adult parasite take hold. Migration of a single larva is therefore limited in time; most die within a month, although longer survival is possible (more than 6 months).

There are also cases of "larva currens" induced by larvae of *Strongyloides stercoralis* (which from the rectum then penetrate the surrounding skin), in which the speed of progression is 10 cm per day [37].

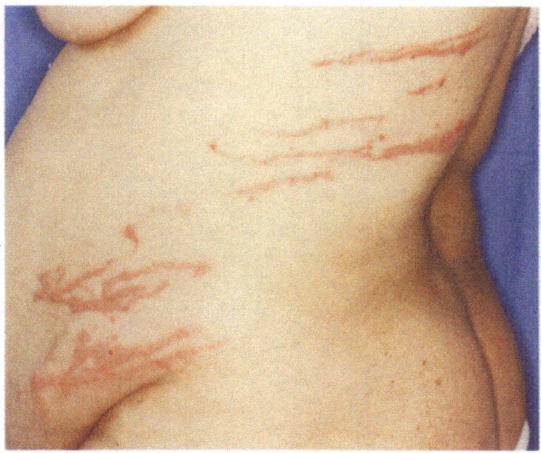

Fig. 9.8. Larva migrans cutanea

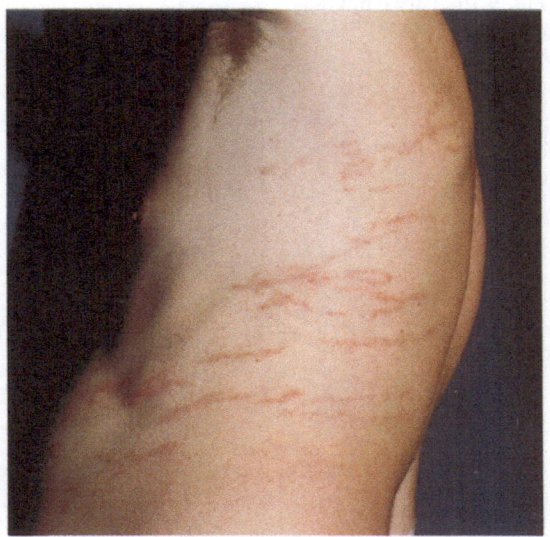

Fig. 9.9. Larva migrans cutanea.
Reproduced with permission
from [39]

The skin sites most commonly affected are those which come into contact with the earth, i.e. the feet, hands and buttocks. Localization in the oral mucosa has also been reported [38]. Damp and sandy terrain, contaminated by the faeces of cats and dogs, is the ideal environment for maturation of the eggs and survival of the larvae. This disease most commonly affects bathers, agricultural workers, gardeners and sewage-farm workers.

The incubation time varies and may last a few months. In most cases, 24-48 h after penetration of the larvae, the entry site shows an erythemato-papular lesion with one or more tunnel-like formations radiating out from it. The linear lesion is about 2-4 mm wide, of a variably intense reddish colour, and only slightly raised in comparison with the surrounding skin (Figs. 9.8-9.12). Itching, pain and pustular lesions are sometimes also present.

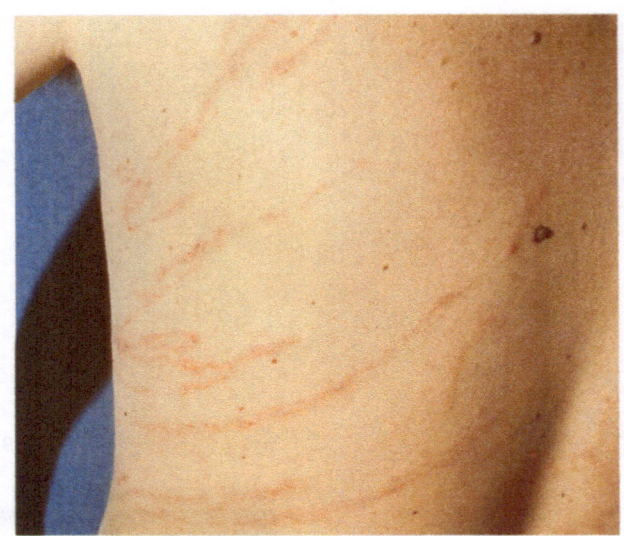

Fig. 9.10. Larva migrans cutanea

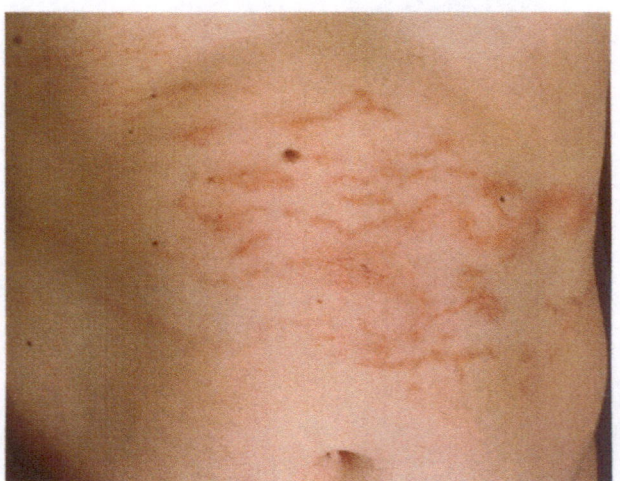

Fig. 9.11. Larva migrans cutanea. Reproduced with permission from [39]

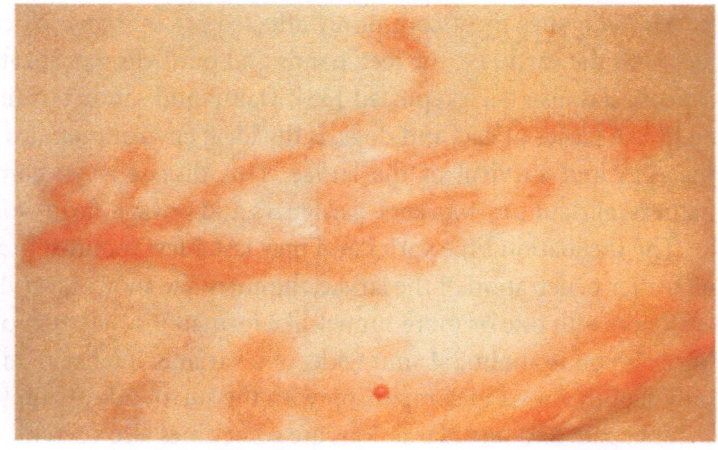

Fig. 9.12. Larva migrans cutanea. Reproduced with permission from [18]

The affliction is self-healing. The larvae stay in the skin for a time ranging from 10 days to 55 weeks, as reported in a review of 96 cases of creeping eruption [40]. Some larvae can also remain quiescent for a while before resuming their migration process [41].

Histological findings are generally aspecific, and the larvae are very difficult to isolate because they progress and are very often situated beyond the visible lesion. The derma shows an aspecific inflammation and the presence of eosinophils. An increase in circulating IgE can sometimes be demonstrated, together with a greater or lesser degree of eospinophilia [42]. Some cases featuring an eosinophilic pulmonary infiltrate (Loeffler's syndrome) have been observed [43] and one case of papulous eosinophilic folliculitis [44]. Immunological investigations to search for specific IgG in the patient's serum have now become an important diagnostic tool [33-38, 40-45].

Cryotherapy with liquid nitrogen or carbon dioxide snow is usually efficacious in very mild infestations, but it must be performed over a fairly wide zone beyond the site of progression apparent from the visible lesion. Good results are obtained using a 10%-15% suspension of topical thiabendazole (an imidazolic worm-killer) in eucerine, 3 times a day for at least 15 days, or a 2% suspension in dimethylsulphoxide [42]. Slight scarification of the lesions and the application of thiabendazole and occlusive bandaging can enhance the efficacy of the treatment. Thiabendazole can also be taken orally, especially in cases with diffuse manifestations due to the frequent side-effects (nausea, vomiting, vertigo), at doses of 15-25 mg/kg/die for 7-10 days [35, 46]. Albendazole at doses of 400 mg/die for 3 days is a valid alternative. The latter drug can also be administered in a single dose (400 mg), although this is not always successful. Another possible treatment is a single dose of 12 mg of ivermectin, that heals in 80% to 100% of cases [47].

Onchocerciasis

Onchocercosis, or onchocerciasis, is an infection induced by *Onchocerca volvulus*, a filament-like nematode that is a parasite only of man and the gorilla. The adult worms live in the derma and subcutaneous tissue; some are free while others collect in masses surrounded by a fibrous capsule (onchocercomata). The disease is transmitted by female Diptera of the *Simulium* genus (black fly), that deposit their eggs on plants and rocks washed with rapid-flowing water (streams, rivers, waterfalls). Mobile larvae hatch from these eggs and live in the running water, transforming into adult insects. Simuliidae are infected when they sting a parasitised individual and then pass the larvae on to another host, again through the sting mechanism.

Onchocercosis is present in sub-Saharan Africa (Senegal, the Congo), Saudi Arabia, Yemen and Central and Southern America (Mexico, Guatemala, Venezuela, Colombia and Brazil).

The organs affected are the skin and eyes [48-55]. Skin manifestations start with intense, widespread itching, that can persist for a long time, caused by the migration of the microfilariae and lysis of the adult worms. After an incubation period of 3-36 months, an acute, pruriginous erythemato-papulous exanthema appears (with lesions 1-3 mm in diameter) in various sites: the trunk and lower limbs in the African form, lower limbs in Saudi Arabia and Yemen, the head and chest in the American form. A chronic phase of diffuse lichenification follows which may take on a hypertrophic and verrucous aspect. In a further phase, onchodermatitis presents with hypotrophy or atrophy and hypo- and achromic lesions, producing the typical "leopard-skin" appearance. This delayed skin complaint manifests with onchocercomata, single or multiple nodular lesions generally between 2 cm and 5 cm in diameter, that are movable against the underlying skin layer but do not evolve into ulcers or suppuration.

Ocular lesions (conjunctivitis, irreversible keratitis, uveitis, iridocyclitis, chorioretinitis, blindness) are observed in cases with involvement of the head or after a long-standing infection (10-15 years).

Massive lymphadenopathy is also possible. Laboratory findings reveal eosinophilia, increased total IgE and ESR. Histological investigations demonstrate microfilariae lying lengthwise or winding through the papillary and superficial derma among the collagen fibres. Microfilariae can be seen also in the epidermis and above all in the onchocercomata, where adult worms are also present. Among the recommended immunological tests, it is important to search for direct antibodies to the specific antigen OV-16: these are present in the circulation before the microfilariae in the derma become apparent.

Treatment is with ivermectin (a single dose of 150-200 µg/kg per os). The treatment can be repeated after 6-12 months. Onchocercomata must be surgically removed.

References

1. Ghiretti F, Cariello L (1984) Gli animali marini velenosi e le loro tossine. Piccin, Padova, 67
2. Johnson PG, Vittor BA (1982) Segmented worms. In: Kaplan EH (ed) Coral reefs. Peterson Field Guides. Houghton Mifflin Company, Boston, 134
3. Hashimoto Y, Okaichi T (1960) Some chemical properties of nereistoxin. Ann NY Acad Sci 90:667
4. Okaichi T, Hashimoto Y (1962) The structure of nereistoxin. Agr Biol Chem 28:224
5. Bracq Z (1937) L'"amphiporine" et la "némertine", poisons des vers némertiens. Arch Intern Physiol 44:190

6. Fisher AA (1978) Atlas of aquatic dermatology. Grune and Stratton, New York, 59
7. Chu GWT (1958) Pacific area distribution of freshwater and marine cercarial dermatitis. Pacific Sci 12:229
8. Wood MG, Srolovitz H, Schetman D (1976) Schistosomiasis: paraplegia and ectopic skin lesions as admission symptoms. Arch Dermatol 112:690
9. Hoeffler DF (1977) "Swimmers' itch" (cercarial dermatitis). Cutis 19:461
10. Meneghini CL, Angelini G (1981) Dermatosi professionali. In: Sartorelli E (ed) Trattato di medicina del lavoro. Vol II. Piccin, Padova, 987
11. Gianotti F, Invoni R (1958) La patologia cutanea degli addetti alla monda e al trapianto del riso. Studio eziopatogenetico con particolare riguardo ai rilievi parassitologici dell'ambiente. G Ital Dermatol 99:377
12. Foti C, Bonamonte D, Vena GA et al (2000) Dermatiti da sanguisughe. In: Veraldi S, Caputo R (eds). Dermatologia di importazione. Poletto Editore, Milano, 214
13. Dejobert Y, Martin P, Thomas P et al (1991) Contact dermatitis from topical leech extract. Contact Dermatitis 24:366
14. Nehili M, Ilk C, Mehlhorn H et al (1994) Experiments on the possible role of leeches as vectors of animal and human pathogens: a light and electron microscopy study. Parasitol Res 80:277
15. Haycox CL, Odland PB, Coltrera MD et al (1995) Indications and complications of medicinal leech therapy. J Am Acad Dermatol 33:1053
16. Halstead BW (1992) Dangerous aquatic animals of the world: a color atlas. The Darwin Press Inc, Princeton, 49
17. Di Napoli PL (1988) Patologia da contatto con fauna marina. Stampasma 5:3
18. Angelini G, Vena GA (1997) Dermatologia professionale e ambientale. Vol I. ISED, Brescia, 202
19. Altamura BM, Introna F, Rositani L (1981) Lesività da fauna marina mediterranea. Med Leg 3:13
20. Angelini G, Giglio G, Filotico R et al (1989) Dermatite da contatto con *Nereis diversicolor*. In: Ayala F, Balato N (eds) Dermatologia in posters. Cilag S.p.A., Napoli
21. Montel RL, Gouyer E (1957) L'escavenite. Bull Soc Franc Derm Syph 64:672
22. Moureaux P (1986) Dermite de contact aux protéines animales (à propos de 2 cas). La Lettre du G.E.R.D.A. 3:73
23. Baran R (1987) Dermite des pêcheurs-amateurs. La Lettre du G.E.R.D.A. 4:27
24. Strani GF, Tomidei M, Sartoris S et al (1987) Dermatosi di raro riscontro indotte da attività sportive. Chronica Dermatol 18:725
25. Romaguera C, Grimalt F, Vilaplana J et al (1986) Protein contact dermatitis. Contact Dermatitis 14:184
26. Bardazzi F, Trevisi P (2000) Larva migrans cutanea. In: Veraldi S, Caputo R (eds) Dermatologia di importazione. Poletto Editore, Milano, 182
27. Georgiev VS (2000) Necatoriasis: treatment and developmental therapeutics. Expert Opin Investig Drugs 9:1065
28. Taniguchi Y, Ando K, Sugimoto K et al (1999) Creeping eruption due to *Gnathostoma hispidum* – one way to find the causative parasite with artificial digestion method. Int J Dermatol 38:873
29. Cavalieri R (1977) Creeping disease. Chronica Dermatol 8:107
30. Argenziano G, Satriano RA (1984) Creeping disease. Chronica Dermatol 15:651
31. Raimondo U, Delfino M, Ayala F et al (1984) Dermatosi da larva migrans. Ann It Dermatol Clin Sperim 38:347
32. Loi R, Lecis AR, Figus V et al (1988) Indagini parassitologiche su un caso autoctono di dermatite serpiginosa. G Ital Dermatol Venereol 123:639
33. Di Carlo A, Leone G, Genchi C et al (1989) Utilità dell'indagine immunologica nella "creeping disease". A proposito di un caso ad insorgenza autoctona. G Ital Dermatol Venereol 124:89

34. Chinni L, Stella P, Papi M et al (1989) Larva migrans cutanea: descrizione di 3 casi. Chronica Dermatol 20:306
35. Stromberg BE, Christie AD (1976) Creeping eruption and thiabendazole. Int J Dermatol 15:355
36. Beaver PC (1956) Larva migrans: a review. Exp Parasitol 5:557
37. Orecchia G, Pazzaglia A, Scaglia M et al (1985) Larva currens following systemic steroid therapy in a case of strongyloidiasis. Dermatologica 171:366
38. André J, Bernard M, Ledoux M et al (1988) Larva migrans of the oral mucosa. Dermatologica 176:296
39. Angelini G, Vena GA (2001) Dermatosi acquageniche. In: Giannetti A (ed) Trattato di dermatologia. Vol II. Piccin, Padova, cap. 45
40. Loewenthal LJA, Leeming JAL (1969) Cutaneous larva migrans. Essay on tropical dermatology. Excerpta Medica Foundation, Amsterdam
41. Stone OJ, Willis CJ (1967) Cutaneous hookworm reservoir. J Invest Dermatol 49:237
42. Kahn G, Johnson JA (1971) Serum IgE levels in cutaneous larva migrans. Int J Dermatol 10:201
43. Guill MA, Odam RB (1978) Larva migrans complicated by Loeffler's syndrome. Arch Dermatol 114:1525
44. Czarnetzki BM, Springorum M (1982) Larva migrans with eosinophilic papular folliculitis. Dermatologica 164:36
45. Higashi GI (1984) A review: immunodiagnostic tests for protozoan and helminthic infections. Diagn Immunology 2:2
46. Stone OJ, Mullins JF (1963) First use of thiabendazole in creeping eruption. Tex Rep Biol Med 21:422
47. Caumes E (2000) Treatment of cutaneous larva migrans. Clin Infect Dis 30:811
48. Elgart ML (1989) Onchocerciasis and dracunculosis. Dermatol Clin 7:323
49. Lombardo M, Girolomoni G, Pincelli C (1993) Oncocercosi oculo-cutanea. Descrizione di un caso. G Ital Dermatol Venereol 128:541
50. Maso MJ, Kapila R, Schwartz RA et al (1987) Cutaneous onchocerciasis. Int J Dermatol 26:593
51. Murdoch ME, Hay RJ, Ramnarain N et al (1990) A clinical classification and grading system of the changes in onchocerciasis and histopathological findings. Br J Dermatol 123 [Suppl 37]:28
52. Poltera AA, Reyna O, Zea Flores G et al (1987) Detection of skin nodules in onchocerciasis by ultrasound scans. Lancet I:505
53. Walter M, Podda M (1993) Oncocercosi. Quaderni Istopatol Dermatol 11:41
54. Yarzabal L (1985) The immunology of onchocerciasis. Int J Dermatol 24:349
55. Veraldi S (2000) Oncocercosi. In: Veraldi S, Caputo R (eds) Dermatologia di importazione. Poletto Editore, Milano, 203

10 Dermatitis caused by fish

Fig. 10.1. *Scorpaena notata* (scorpionfish)

There are more than 500 species of vertebrate fishes that are poisonous to man, while another 250 species have venomous organs or apparatus that can cause very painful, sometimes fatal wounds. Both poisonous (cryptotoxic) and venomous (fanerotoxic) (*faneros* = clear, evident) fish pose a serious economic and social problem, as the former are a handicap to full exploitation of marine resources, while the spines and venomous sting apparatus of the latter are a danger to bathers, fishermen and other people engaged in water activities. The problem is aggravated by the fact that even on the rare occasions when the biotoxin can be identified, there are no known remedies able to neutralize the poisoning symptoms.

Passively toxic fish

Poisonous fish are also known as passively toxic fish; in other words, they do not produce toxins for defence or offence purposes but acquire them from the environment where they live, and are not therefore poisonous in themselves but only when eaten. In these cases, the exogenous origin of the toxin has not always been ascertained. Moreover, strangely enough in the same environmental conditions some species are poisonous while others are not, or else some individuals of the same species may be poisonous and others not, or again species living around a particular island, for instance, may be poisonous whereas others of the same species living near a different island a few miles away may not. In any case, these kinds of fish are known to populate the tropical regions above all.

The best known ichthyotoxins studied up to now are ciguatoxin and tetrodotoxin. The former does not induce immunity and probably derives from unicellular algae, especially Cyanophyceae and Dinoflagellates, the first link of the food chain of carnivorous fish. The latter has been isolated in pufferfish, otherwise known as blowfish (so-called because they blow up with water or air when molested), porcupinefish (that can inflate and whose whole body is covered with long, hard spines) and sunfish (that have a flattened, ellipsoid body), all of which belong to families that live in the waters around Japan and South-East Asia.

Human food poisoning caused by fish can also be a concern in non tropical countries owing to imports of frozen fish from tropical zones. The viscera of these fish are generally more toxic than their flesh: the most toxic organs are the liver, intestine and gonads, in decreasing order. For this reason, some populations of the Pacific remove the viscera immediately when harvesting these animals. In the field of study of marine animal biotoxins, no chemical methods of analysis have yet been found for establishing toxicity. Food poisoning from passively toxic fish is very often fatal [1].

Actively toxic fish

Venomous fish have specialized organs, a venomous apparatus or glandular structures that can secrete toxic substances. These can be inoculated by means of a sting or a bite, and are intended to paralyse the prey. Man can inadvertently become the victim of these animals either in the water (mainly as a result of a defensive reaction and only very rarely of a direct attack) or outside, due to wrong handling of these fish [2-6].

There are various species of venomous fish, belonging to two classes, those with a cartilaginous structure, the Chondrichthyes, and those with a bony skeleton, the Osteichthyes (Table 10.1) (Fig. 10.1) [1-11].

Table 10.1. Fish with toxic characteristics present in the Mediterranean

1. Class: Chondrichthyes	2. Class: Osteichthyes
Order: Batoidea or Rajiforme (stingrays or trygonidae)	A. Family: Trachinidae (weeverfish) Species: *Trachinus araneus* (weeverfish) *Trachinus vipera* (lesser weeverfish) *Trachinus draco* (greater weeverfish) *Trachinus radiatus* (radial weeverfish)
A. Family: Dasyatidae Species: *Dasyatis pastinaca* (common stingray) *Dasyatis violacea* (purple stingray) *Dasyatis centroura* (thorny stingray)	
	B. Family: Scorpaenidae (scorpionfish) Species: *Scorpaena porcus* (black sea pig) *Scorpaena scrofa* (large or red scorpionfish) *Scorpaena ustulata* (small scorpionfish) *Scorpaena dactyloptera* (seabottom scorpionfish)
B. Family: Myliobatidae	
	C. Family: Muraenidae (moray eels) Species: *Muraena helena*

Fig. 10.2. Ray or trygonida

The Chondrichthyes class

The Chondrichthyes are one of the largest and most important groups of toxic marine organisms (second only to the Coelenterates in terms of the number of human cases of poisoning inflicted by their venoms). These are known by the generic name of stingrays, and have a flat body of rhomboid or trigonal shape (hence their other name trygonidae) and a long, whip-like tail with one or more sharp spines on the dorsal surface (Fig. 10.2). It has been estimated that there may be upwards of 1,500 stingray attacks per year in the United States alone. Most rays are seawater animals and only one family, the Potamotrygonidae, lives in torrents or rivers in South America, Equatorial Africa and Laos. There are 6 different families of marine rays.

They are common inhabitants of tropical, subtropical and temperate seas, and vary greatly in size, ranging from 6 cm to 6 m or even more. They generally live on sandy or muddy bottoms, rarely deeper than 35 m, and have no migratory habit. Thanks to their flat bodies, rays pass long periods of time in shallow waters, burrowed half-hidden under a thin layer of sand with only their eyes protruding.

The venomous apparatus consists of a caudal, conical spine situated in the proximal part of the tail. It has the shape of a saw and a length ranging from 4-6 cm in Mediterranean species to over 40 cm in tropical species. Inside the spine, there are glandular structures and ducts that serve to secrete and inoculate the toxic substances. The caudal spine is caduceus: after formation of a new spine, the old spine is shed when the new one has grown to the same length. There are four different types of venomous apparatus; the one described belongs to the *Dasyatis* species and is notoriously the most dangerous. The spine is made of a hard material similar to bone, called vaodentine.

Fig. 10.3. Treading on the body of the ray causes violent projection of the spine against the victim's leg

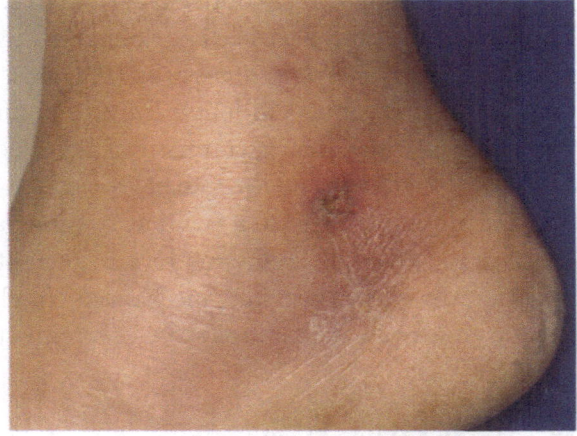

Fig. 10.4. Wound from stingray spine. Reproduced with permission from [5]

Most stingray injuries are due to inadvertently stepping on the body of the ray: pressure in this zone triggers a defensive mechanism whereby the fish arches its tail and projects the spine violently against the foot or ankle (Fig. 10.3). Exceptionally, if the victim is lying in the sand, the spine may strike the chest and in this case the wound may be fatal. Stingray wounds feature a laceration or puncture of variable depth, and on penetration of the skin, the toxic substances are released into the underlying tissue (Fig. 10.4). The damage is thus both mechanical and chemical. The venom, contained in two ventrolateral grooves on the underside of the spine and produced by holocrine glandular cells, consists mainly of protein structures with a high molecular weight. The pure extract contains serotonin, 5-nucleotidase and phosphodiesterase. The venom acts on the cardiovascular system (inducing vasodilatation or vasoconstriction according to the concentration, atrioventricular blockage, cardiac arrest), the respiratory tract (depressing the respiratory centres) and the neuro-logical system (causing convulsions).

Wounds inflicted by rays, unlike those caused by other venomous fish, are very wide and have jagged edges, sometimes requiring surgical treatment: a spine 4-5 cm long can inflict a wound 20-25 cm long. At the site of the lesion, initial vasoconstriction is followed by intense erythema and in severe cases by perifocal punctiform necrosis. Local pain generally onsets immediately or within 10 minutes of the attack. It is disproportionately severe in comparison with the size of the wound, variously described as acute, throbbing or piercing, and can last 1-2 days. There may be various general symptoms (hypotension, sweating, vomiting, diarrhoea, tachycardia, muscular paralysis), that in very severe cases can even be fatal.

The best known, notoriously dangerous rays of the Dasyatidae family include the *Dasyatis brevicaudata* species (the largest ray in the world, that can attain a length of 4.5 m or more, a width of 2.2 m and weigh over 325 kg; many fatal inflictions have been attributed to this ray, that lives in the Indo-Pacific region), *Dasyatis dipterurus* (the most common ray in Central America, that is about 2 m long), *Dasyatis pastinaca* (the European ray, or "pastinaca marina", that is common in the Mediterranean, the north-east coasts of the Atlantic and the Indian Ocean, and can attain 2.5 m in length) and *Dasyatis americana* (that is about 2 m long and lives along the western coasts of the Atlantic). Freshwater rays are present largely in the Amazon Rio and Rio de la Plata. Their sting mechanism, the clinical symptoms and treatment measures are the same as those of marine stingray attacks [12, 13].

The Osteichthyes class

The Trachinidae family

Trachinidae or weeverfish are relatively small fish (from *Trachinus vipera*, at about 10-12 cm to *Trachinus draco* at 40-50 cm) (Fig. 10.5) of the Teleostei family, present along all the Mediterranean coasts and the North-East Atlantic (European coasts). All the species have spines on various parts of the body linked to cells that secrete a venom similar to snake-venom. The Latin name derives from the Greek *traknos* (stinging). In Italy these fish are also known as "pesce ragno" (spiderfish).

Weeverfish are among the most dangerous venomous fishes in temperate zones, and live in shallow waters or on the shore, half-buried in the sand with only the dorsal spine exposed. They have chameleon-like characteristics and blend with their surroundings. Weeverfish are sluggish and carnivorous

Fig. 10.5. *Trachinus draco* (weeverfish)

Fig. 10.6. Treading on a weeverfish

and can survive out of water for a time; even after death their spines are venomous.

They have a dual venom apparatus consisting of two opercular spines and 5-8 dorsal spines. The glandular tissue that secretes the venom is present at the base of the spines, while the excretory tubules terminate on the surface of the spine; the relative orifices are covered with a fine sheath to prevent release of the venom. When threatened, weeverfish blend even further with the background colour and remain immobile, only arching the dorsal spines for defensive purposes.

Accidental contact with a weeverfish is very common along Mediterranean shores, generally due to inadvertently treading on one (Fig. 10.6).

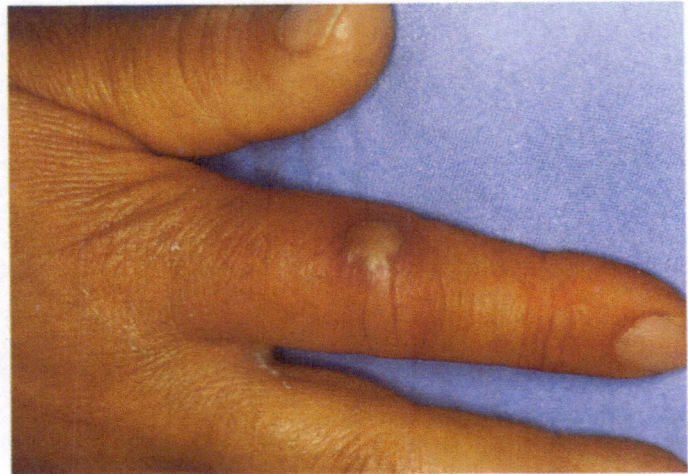

Fig. 10.7. Stingray lesion. Reproduced with permission from [5]

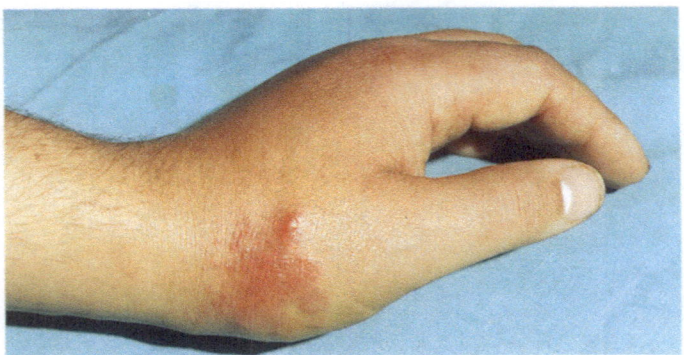

Fig. 10.8. Stingray lesion. Reproduced with permission from [5]

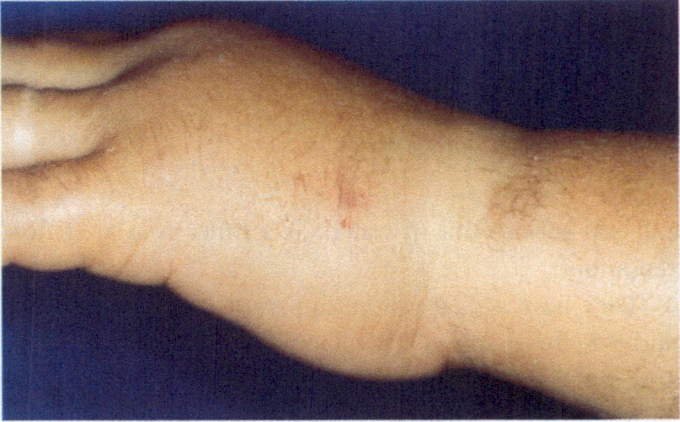

Fig. 10.9. Stingray lesion with intense oedema of the arm. Reproduced with permission from [5]

Fishermen can also be stung, while they are removing fish from their nets or lines (Figs. 10.7-10.9), as well as skin divers, since the spines are very strong and can pierce gloves and flippers. The chemical nature of the venom is little known; it has a strong anti-cholinesterase action and toxic activity on various

Fig. 10.10. *Scorpaena porcus* (black scorpionfish)

organs, especially the heart. The rare fatalities are caused by cardiac arrest. Local pain is disproportionately severe, lasts 16-24 hours and irradiates around the limb.

There are 4 known species of weeverfish, among which *Trachinus draco*, that can attain a length of 45 cm, is widespread from Norway down to the Mediterranean and along the coasts of North Africa, and *Trachinus vipera* lives in the North Sea and the Mediterranean and is 15 cm long.

The Scorpaenidae family

Scorpaenidae, or scorpionfish, belong to the most numerous and geographically widespread family of toxic fishes, being present in all the warm and temperate seas on earth and even in the cold Arctic Sea. There are about 80 species with a venomous defensive and offensive apparatus. They are all shallow-water bottom dwellers, and have a short, brilliantly coloured body, very large eyes and mouth and large fins. Some of them can blend perfectly into the surrounding environment.

Toxic Scorpaenidae are subdivided into 3 groups, according to the shape of the venom organs: scorpionfish (*Scorpaena*), zebrafish (*Pterois*) and stonefish (*Synanceja*).

Scorpionfish live in bays, along sandy beaches, rocky coasts or coral reefs, from the tidal zone down to a depth of about 150 m. About ten different species have been reported to be endemic to the Mediterranean; the most frequently encountered are *Scorpaena porcus* (Fig. 10.10), a dark coloured

Fig. 10.11. *Scorpaena scrofa* (red scorpionfish)

fish about 20-30 cm long and *Scorpaena scrofa* (Fig. 10.11), that has a typically red skin and attains a maximum length of 50 cm. Other species known to be particularly toxic include *Apistus carinatus* (common along the coasts of India, Indonesia, the Philippines, China, Japan and Australia, this species is about 16 cm long), *Scorpaena guttata* (common along the Californian coasts) [14], *Scorpaena plumieri* (present on the Atlantic coast from Massachusetts to Brazil) and *Scorpaena mystes* (that lives along the coasts from Mexico to Peru).

The venom apparatus of these species consists of short, thick spines: 12 dorsal, 3 anal and 2 pelvic. The glandular tissue and tubular secretion structures are situated within the grooves of these spines. The tubular orifices are again near the points of the spines and enveloped in a thin sheath. This membrane ruptures on contact, enabling inoculation of the venom, a heat-labile protein compound with a high molecular weight and a serotonin-like effect.

Scorpionfish spend long periods of time on the sea bottom, preferring rocky, irregular bottoms. Human contact with scorpionfish is generally accidental during fishing expeditions by boat or underwater, especially during harvesting procedures.

Scorpaenidae of the *Pterois* group (zebrafish), present in the Pacific, from the east coast of Africa to India, the Philippines and Australia, are among the most beautiful fish in existence. They have very long, feathery pectoral fins and lacy dorsal fins. As they swim, they wave their fins gently and look like birds in flight. They seem innocuous and inviting but one must resist the temptation to touch them: hidden under the lacy and feather-like fins are the slender venom spines. The venom apparatus is similar in shape and location to that of scorpionfish. The best known

Fig. 10.12. *Muraena helena*

species are *Brachirus zebra* (they are 12 cm long and live in regions of the Indian and Pacific Oceans), *Pterois antennata* (30 cm long, present in the Indian and Pacific Oceans) and *Pterois volitans* (one of the most beautiful, spectacular fish among the coral reefs, to be admired from a distance: beware of contact!).

Scorpaenidae of the *Synanceja* type (stonefish) are commonly found in tidal pools and shoal reefs, lying motionless among the corals, under rocks or hidden in the sand or mud. They appear entirely uninterested in careless human intrusions. They have the same number of spines in the same sites as the other Scorpaenidae, but these are short and thick and the venom apparatus is highly developed. Most of these species are widespread in Australia (*Inimicus barbatus*, 20 cm long), Japan (*Inimicus japonicus*, 20 cm long), Indian and Pacific Oceans (*Chloridactylus multibarbis*, 10 cm long; *Synanceja horrida*, an extremely dangerous fish about 60 cm long; *Synanceja verrucosa*, 30 cm long, also present in the Red Sea).

Stings from Scorpaenidae, whatever the species or type, cause such fierce, anguishing pain that the victim may lose consciousness. The general symptoms are much more serious after stings from the *Pterois* and *Synanceja* species than from *Scorpaena*, but stings from the latter type are more frequent. Some fatal cases of stings from this genus have been reported.

The Muraenidae family

There is still debate as to the toxicity of the Muraenidae family, and especially the most widespread species in the Atlantic Ocean and Mediterranean, *Muraena helena* (Fig. 10.12). Moray eels have a long, cylindrical body flattened posteriorly, and a small head with a wide buccal aperture. They usually live on rocky sea bottoms at depths of over 15 m. They have no spines on their bodies but their powerful jaws are equipped with a row of sharp, strong teeth, unconnected with glandular or secreting substances. The venom is in fact produced by unicellular glands lining the dome of the palate and is present only in the blood (haemoichthyotoxin). The toxin may be injected into the wound inflicted by a bite through contact with toxic saliva or by direct compression against the animal's palate.

The same type of blood toxin is present in all the Eel order, consisting of *Anguilla, Muraena* and *Conger.* This ichthyotoxin is a heat-labile protein and is not therefore harmful to man because it is inactivated by cooking. However, one must be careful to avoid lacerating wounds when handling the eel.

The traditional view that moray eels are voracious, vicious animals quick to attack is a myth, and probably derives from the fierce appearance they assume as they repeatedly open their jaws wide to breathe. This sight, especially in an animal that has attained 1.5-2 m length, can cause a diver to panic and thrash about frantically, which may be very dangerous because bigger moray eels tend to nestle inside grottoes and crevices, where the diver's movements should be carefully controlled.

Contact with a moray eel generally occurs while capturing and handling it. Moray eels are very muscular, resistant animals and do not surrender without fighting strenuously. When harpooned, the eel winds around the barrel of the weapon and tries to bite anything within reach. The battle between fisherman and eel is particularly difficult when long periods of decompression have to be observed. A bite from a moray eel, especially a large one that was well-anchored within its lair, sometimes prevents a diver from being able to resurface, resulting in very dangerous, sometimes fatal consequences. Once the eel has been pulled on board the boat, it does not die quickly and so it is necessary to kill it fast to avoid furious bites, particularly as its slimy hide is very slippery.

The local pain from a bite is tearing and intense and may be associated with general symptoms such as breathlessness and collapse. No fatal cases have been described in the literature.

The venom present in the blood that makes the eel poisonous and venomous is fairly complex and has not yet been fully identified. It contains proteolytic enzymes (hyaluronidase, substances activating bradyquinines), haemotoxic (with a haemolytic action), neurotoxic and cytotoxic substances with a local action (curaro-like paralysing toxins) and small peptides with a hypotensive action.

Clinical symptoms

The clinical symptoms of poisoning due to stings or bites from the various actively toxic fish are much the same; the severity of the signs is generally proportional to the type of venom and to the quantity inoculated. In any case, the appearance of the wound is no guide to the severity of the clinical picture.

The pain is instantaneous, constant and extremely fierce. It is worst in the case of weeverfish stings and to a lesser degree, those from rays or scorpionfish and only slightly less intense after a bite from a moray eel. The pain irradiates around the whole limb within 30 minutes and persists unabated for 12-48 hours. This is generally followed by further intermittent paroxysms. In all cases it is so violent as to provoke malaise, lipothymia and loss of function of the limb. Clearly, all these symptoms can seriously impair the underwater diver's ability to emerge correctly.

The wound generally appears slight and falsely reassuring. The type of wound can help to identify the exact species involved. An extensive, lacerated wound suggests a stingray, and is most frequent on the legs or arches of the feet, whereas a pointed, copiously bleeding wound localized on the feet or hands is likely caused by a weeverfish.

At the moment of the sting or soon after, a severe inflammatory reaction will onset, with ischaemic necrosis, pallor, cyanosis, purpura and blood-blisters. Extensive, hard and painful oedema will then ensue, followed by lymphangitis and later by satellite adenopathy. However, the signs of inflammation are often masked by the tissue ischaemia.

The tearing pain induces anguish, tachycardia, dyspnoea and hypotension. These signs of shock may be very pronounced, and even lead to syncope and death. Neurological symptoms are sometimes present with vertigo, parasthesia, contracture or muscular spasm, convulsions and delirium. Reynaud's phenomenon has also been observed a few weeks after a sting from a weeverfish, on one finger only, at the site of the sting [15].

In the Mediterranean, evolution is always favourable and any risks are linked largely to its effect on underwater swimming activities. The most frequent contact is with weeverfish; an encounter with a stingray is less common whereas contacts with scorpionfish are relatively frequent but have a more benign evolution.

Treatment and prevention

Treatment must be immediate and expert, carried out at the site of the accident. The principal aim is to limit the spread of the poison as much as possible. The wound should be washed with seawater and, if possible, all accessible fragments of spines extracted, without lacerating the tissues. If the wound involves a limb, spread of the poison can be circumscribed by a haemostatic ligature, which must not block arterial flow and must be released for 90-120 seconds every 10 minutes.

Sucking out the poison is not entirely risk-free, despite the fact that the poison is largely inactivated by the digestive juices, and in any case is not very efficacious. If available, a "poison-aspirator" can be used for this purpose. This is an easy-to-use, painless first-aid kit consisting of a syringe-shaped mini-pump that, in a fraction of a second, can provoke a depression that can extract much of the poison and soothe the pain. As the pump is portable and can be applied to all people, it is an essential item in any first-aid kit. However, it is absolutely no substitution for prompt hospital admission for the necessary general and local symptomatic treatment.

To denature and inactivate the venom (which is heat-labile and is destroyed at 50° C), the affected part (generally a limb) should be immersed in water as hot as possible, to which a disinfectant may be added. Soaking should be continued for 1-2 hours. As hot water is not always available on a beach, as an alternative the burning tip of one or two cigarettes can be applied for 5-10 minutes, if the patient can bear it. At the site of the event, the general treatment must aim to soothe the pain. Local injections of 1% xylocaine without adrenaline can be given, or intravenous calcium gluconate, potent analgesics or morphine, tranquillizers. Corticosteroids are useful to relieve the symptoms of both shock and the poison.

In practice, administration of antihistamines does not seem to be very effective, while no sensitisation mechanisms seem to be triggered by contact with fish venoms and so no signs of anaphylaxis are observed.

Once in hospital, the wound will be cleaned, incised and debrided to remove spines, and if necessary, surgical sutures will be applied. The patient may need rehydrating, oxygenation and cardio-circulatory intensive care. Analgesic drugs will be administered for a long period. In general, the conditions do not justify the use of heparin.

Prevention is by avoiding the temptation to walk barefoot in shallow waters, and istead wearing plastic sandals and above all, a mask, wetsuit and fins for underwater diving. These fish should never be picked up by hand, even if they are lying on the beach or the bottom of the boat, as the spines remain poisonous for many hours after death. There is no danger at the fishmonger's, because there is a legal obligation to remove the spines and stings of venomous fishes before displaying them for sale. But if in any doubt, better safe than sorry! Instead, owing to the fact that the venom is heat-labile, a meal of these fish can be enjoyed in all safety.

Catfish stings

Catfish live mainly in freshwater in Eurasia, North and South America, Africa and Australia. They belong to the super-order of Ostariophysia and the order of Siluriformes. Most of these fish (30 families and about 2400 species) live in tropical rivers and streams; some live in temperate zones (Icthaluridae, Siluridae, Diplomystidae, Bagridae), while only two families, Plotosidae and Ariidae, live in the sea [16]. They are called catfish because of the long, fleshy barbels that grow from their snout. There are various shapes and sizes of catfish: in South America, especially in the Andes and Amazon rivers, the species are very tiny (13 mm in Bolivia), whereas in Amazonas they may attain 3 m.

Some fishes of the Pygidiidae family are "parasites" of bigger catfish, living and depositing their eggs in the gills. These strange fish, present only in South America in the rivers of the Andes, have no scales and are slender, transparent or brownish, varying in length from 3 mm to 5 cm. They are unusual in having an elective tropism for mammalian urine, including human urine. They swim against the flow of urine and enter the urethra, the vagina and rectum while the mammal is urinating into the water. These fish (called candirú or carnero by the natives of South America) probably mistake the flow of urine for the jet of water expelled from the gills of large catfish. The best known species is *Urinophylus erythrurus*, so-called for its urinotropic tendencies and the fact that it feeds off blood [17].

The venom apparatus of catfish consists of a single, robust cutting spine (sometimes with jagged teeth along it) in the dorsal and pectoral fins, and axillary venom glands. The most common marine species are *Galeichthys felis* (in the Gulf of Mexico), *Clarias batrachus* (the Indian Ocean) and *Bagre marinus* (North and South America). *Noturus furiosus* is common in the rivers of North Carolina and *N. miurus* is present in the Mississippi.

The pain from a catfish sting is instantaneous and piercing, and may be localized or irradiate around the whole limb. Some tropical fish (*Plotosus*) provoke violent pain that persists for 2 or more days. The affected part immediately becomes pallid. This pallor is followed by cyanosis and then by erythema and oedema; the latter is very severe in some cases and is associated with torpor and gangrene of the damaged zone. Shock may also ensue. Supervening bacterial infection of the wound is a common problem. Some fatal cases of stings from tropical catfish species have been reported. It should also be borne in mind that catfish are common in aquariums and that it is dangerous to handle them [2, 18, 19].

Soapfish dermatitis

Among actively toxic fish, some produce toxins in the skin glands to repel attacks. Apart from fish, toxic secretions are also produced by Opisthobranch Molluscs and, among land animals, amphibians.

The discovery of toxic skin secretions in fish was made thanks to chance observations of irritations on the hands due to touching these fish, and to having seen that other fish die if placed in the same tank. Many tropical fish belonging to various families are toxic. From the *Ostracion lentiginosus* species of the Ostraciontidae family, a toxin called pahutoxin (from the Hawaiian name "pahu", after the species that produces it) has been isolated [20]. This fish is called boxfish by the Americans (analogously, the Italian name is "pesce cofano"), owing to its box-shaped body. Pahutoxin is a choline ester with β-acetoxypalmital acid and in aqueous solution it foams like saponin. It has a strong haemolytic action on other fish but is not toxic to man, who may, however, develop contact dermatitis.

Other ichthyotoxins from the skin secretions of other fish, although chemically unlike pahutoxin, also have a haemolytic action. There are many toxin-secreting fish of the Grammistidae family, known as soapfish because they produce large quantities of foam when they are put in a tank or otherwise disturbed. This toxin, called grammistine, seems to be a mixture of peptides with tertiary or quaternary amine groups. A well known species found in the Virgin Isles and Puerto Rico is *Rypticus saponaceus,* contact with which induces acute dermatitis together with itching and burning sensations.

Fish equipped with an electrical apparatus

Electric fish are among the most interesting of the harmful aquatic animals. Many plants and animals give off an electric shock, and there are approximately 250 species of fish with electric organs. Electricity is an essential part of the metabolic activity of living creatures. In most cases, however, the amount of current emitted is so small that it can be detected only by means of highly sensitive instruments. In land animals, the air acts as a good insulator and these small discharges are not therefore usually noticed. Instead, water is an excellent conductor, and aquatic animals and some fish possess specialized organs that discharge electricity through the water at very high voltages.

Only a few fish produce electricity as high as 650 volts, such as the electric eel *Electrophorus*; most emit discharges ranging from some millivolts to several volts. High discharges are only emitted in exceptional circumstances as a defence mechanism or to stun their prey, whereas weak discharges are emitted continually as an electrolocation system and for social communication. The electric organs of fishes function through particular cells (electrocytes or electric cells) arranged in series, that are synchronously excited by spinal nerve signals to generate small voltage gradients running in the direction of their linear arrangement. As each stack of electrocytes is hermetically enclosed by insulating tissue, the voltage of each electrocyte contributes linearly to the whole sum, just like a series of small batteries [21].

The electrolocation process resembles that of a sonar: the animal sounds out the surrounding environment by emitting signals and monitoring their feedback. In this way, such animals can operate in total darkness. The fish generates a current field that is triggered by the anterior part of the body and converges at the tail extremity. An object with a different impedance (opposition in an electric circuit) in the surrounding water distorts the electric field and so alters the model of electric current intensity on the body of the fish. In this way, by monitoring the local variations in electric activity, the fish can perceive the nature of surrounding objects.

The best known and most studied of aquatic organisms belong to the Electrophoridae and Torpedinidae families. The latter family has only a single species, *Electrophorus electricus,* an eel that inhabits the streams and swamps of South American jungles. It is the most powerful of the electric fishes and can produce electric shocks up to 650 volts, with an average of 40 watts, enough to light up an electric bulb. When this current is discharged into the water, it sets up an electric field that is sufficient to stun a man or even a horse. At rest, *Electrophorus* gives off no electricity, but when it is swimming it emits a discharge of about 50 volts/sec. It can give out a steady series of discharges for 20 minutes, rest for 5 minutes and then start again: it can truly be said to be one of the most efficient batteries in the world!

The Torpedinidae family, of the Rajiforme order, includes 30 species. These fish have an oval, flattened body and a large, short, strong robust tail. The kidney-shaped electric organs are arranged dorsally in pairs, one on each side, and can emit current up to 200 volts. In emergency situations, torpedo rays give off shocks at a relatively high voltage and amperage. These diminish progressively in power and a rest period is required to restore the starting electric potential. In the past, torpedo rays were used in electrotherapy: the Greeks called these fish "*narke*" (lethargy), from which the terms narcosis and narcotic are derived. Ancient Roman doctors used torpedo rays to inflict an electric shock on their patients.

Among the best known species, *Narcine brasiliensis* lives inshore along the western Atlantic coast (Florida, Texas, Brazil, Argentina), while *Diplobatis armata* and *Torpedo californica* inhabit the coast of southern California.

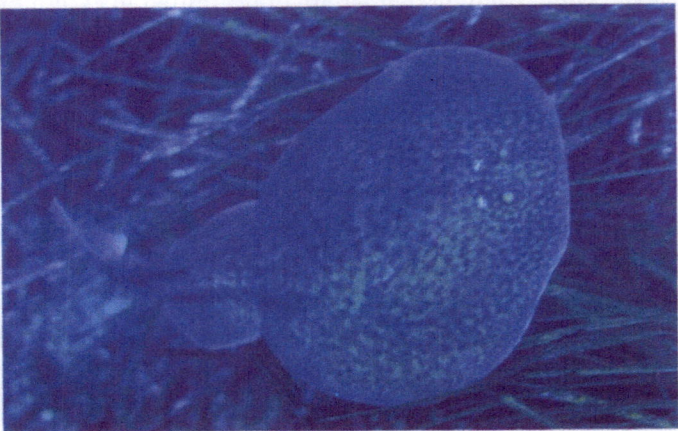

Fig. 10.13. *Torpedo marmorata* (marbled stingray)

Three species of torpedo rays live mostly burrowed under sandy or muddy bottoms in shallow depths in the Mediterranean: *Torpedo torpedo, Torpedo marmorata* (Fig. 10.13) and *Torpedo nobiliana*. The first two attain a maximum length of 60 cm and live along the coast at depths ranging from a few metres to about a hundred. The third species is the largest and can attain 1.80 m and a weight of about 70 kg: this is the most dangerous species for man.

Injury occurs by means of direct or indirect contact during capture and handling of the animal. In fact, if a diver touches his metal rod after having harpooned a torpedo ray, he will get an electric shock more or less proportional in power to the size of the animal. The shock is not generally strong enough to cause direct skin or nerve lesions, apart from a slight state of stupor induced by very large animals. The greatest danger to the diver may be abnormal, incorrect emersion due to loss of awareness or panic caused by the unexpected shock, occurring in an unfavourable environment.

Fish inducing mechanical injury

There are fish that can inflict a mechanical injury on man present in all the seas, although such species are not very numerous. Fish with a cartilaginous skeleton belong to the Rajiforme and Squaliforme orders, and fish with a bony skeleton to the Anguilliforme order.

The noxious characteristics of sharks are only too well known. They are common in tropical and temperate waters and in the oceans at the Poles and dwell at all depths. They have a cartilaginous skeleton and are covered in placoid scales; they have an asymmetrical tail and their liver can weigh up to 20% of their body weight. Their immune system is unequalled and they have a superb sensory system. Most of them are powerful, fast swimmers. There are about 250 species of sharks, and about 27% of them are known to attack man. Attacks have been reported between 46° of latitude north and 47° south, ranging from the Upper Adriatic Sea to southern New Zealand. Most attacks are seasonal and occur in waters where the temperature is above 20° C (68° F), during the warm periods (from May to October in the Northern Hemisphere and from November to April in the Southern Hemisphere), when sharks are more likely to encounter man. The most dangerous time is the evening and night when sharks feed most voraciously.

About 8% of them attain a length of over 2 m but their size is not necessarily proportional to the risk they pose: in fact, despite its large size (it can exceed 13 m), *Rhiniodon typus* feeds off plankton and is not interested in man. However, all species longer than 1.2 m must be considered potentially dangerous to man, especially in the presence of blood or food in the water. In most attacks on man, the shark was not seen prior to the moment of contact. Although there is no uniform pattern of behaviour, sharks swim in an erratic, irregular fashion before attacking: they hunch their backs and extend their pectoral fins forwards, while swishing their head and tail back and forth. Sometimes an attack is preceded by one or more variably violent contacts, ranging from gentle bumps to violent collisions. However, many attacks are the result of directly provoking the animal, by pushing it, trying to grab it by the tail or tantalizing it with dead fish.

Sharks are also present in freshwater lakes (Nicaragua and Izabel in Guatemala, Jamoer in New Guinea) and rivers (Zambezi and the Ganges, the rivers of the Amazon, Australia, Asia, Iraq and South-East Africa).

Giant manta rays, that can be up to 6 m long and weigh up to 1500 kg, feed off plankton and are not aggressive but they can still be harmful owing to their great body mass and rough dermal denticles that can cause abrasions on contact with human skin.

Although barracudas are carnivores, they do not tend to attack man. They are attracted by any shiny object and are widely distributed throughout all tropical and subtropical zones of the world.

References

1. Ghiretti F, Cariello L (1984) Gli animali marini velenosi e le loro tossine. Piccin, Padova, 125
2. Fisher AA (1978) Atlas of aquatic dermatology. Grune and Stratton, New York, 71
3. Kaplan EH (1982) Coral reefs. Peterson Field Guides. Houghton Mifflin Company, Boston, 206

4. Alstead BW (1992) Dangerous aquatic animals of the world: a color atlas. The Darwin Press Inc, Princeton, 77
5. Angelini G, Vena GA (1997) Dermatologia professionale e ambientale. Vol I. ISED, Brescia, 202
6. Angelini G, Bonamonte D (1997) Dermatoses aquatiques méditerranéennes. Nouv Dermatol 16:280
7. Weiller M, Genolier-Weiller A (1987) Accidents cutanés provoqués par la faune sous-marine Méditerranéenne. Première partie. Nouv Dermatol 6:331
8. Weiller M, Genolier-Weiller A (1987) Accidents cutanés provoqués par la faune sous-marine Méditerranéenne. Deuxième partie. Nouv Dermatol 6:354
9. Phillips C, Brady WH (1953) Sea pests, poisonous or harmful sea life of Florida and the West Indies. Univ Miami Press, Miami
10. Mullanney PJ (1970) Treatment of stingray wounds. Clin Toxicol 3:613
11. Russel FE (1971) The stingray: natural history, venom apparatus, chemistry and toxicology, and clinical problem, in: poisonous marine animals. TFH Publications Inc, Neptune NJ
12. Castex MN (1967) Freshwater venomous rays. In: Russel FE, Saunders PR (eds) Animal toxins. International symposium on Animal Toxins. Pergamon Press, New York, 167
13. Rodrigues RJ (1972) Pharmacology of South American freshwater stingray venom (*Potamotrygon motoro*). Trans NY Acad Sci 34:677
14. Schaeffer RC Jr, Carlson RW, Russel FE (1971) Some chemical properties of the venom of the scorpionfish *Scorpaena guttata*. Toxicon 9:69
15. Carducci M, Mussi A, Leone G et al (1996) Raynaud's phenomenon secondary to wee-verfish stings. Arch Dermatol 132:838
16. Banister K, Campbell A (1993) The encyclopedia of aquatic life. Facts on File Inc, New York, 74
17. Alstead BW (1992) Human parasitic catfish (candirú). In: Alstead BW (ed) Dangerous aquatic animals of the world: a color atlas. The Darwin Press Inc, Princeton, 223
18. Scoggin CH (1975) Catfish stings. JAMA 231:176
19. Patten BM (1975) More on catfish stings. JAMA 232:248
20. Boylan DB, Schener PJ (1967) Pahutoxin: a fish poison. Science 155:52
21. Alstead BW (1992) Electric aquatic animals. In: Alstead BW (ed) Dangerous aquatic animals of the world: a color atlas. The Darwin Press Inc, Princeton, 217

11 Dermatitis caused by aquatic bacteria

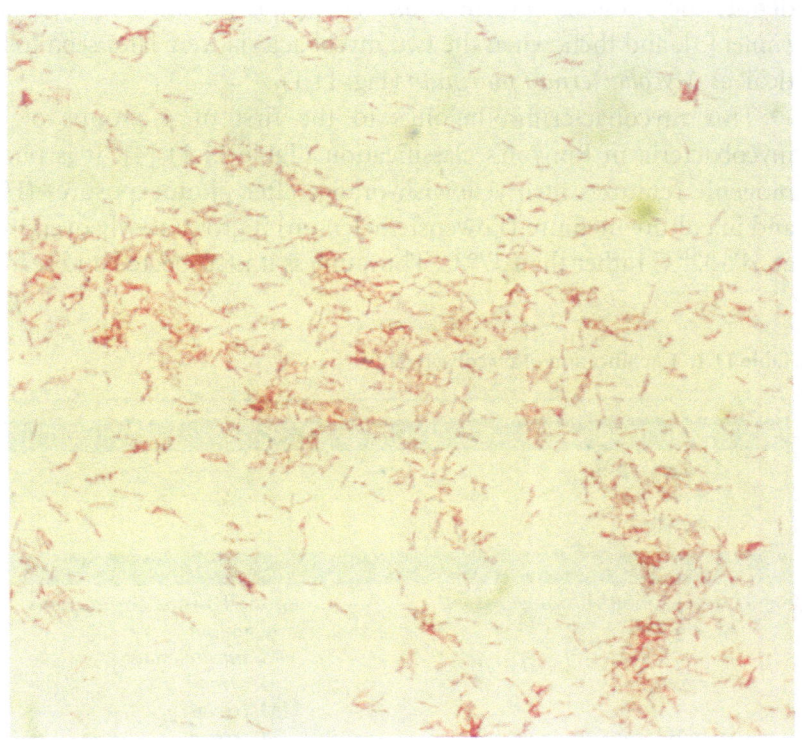

Fig. 11.1. *Mycobacterium marinum*

Infection by *Mycobacterium marinum*

*M*ycobacterium marinum lives in aquatic environments. It was isolated in 1926 by Aronson from tubercles in various organs of marine fish found dead in the Philadelphia Aquarium [1]. This organism was identified as a causal agent of human disease only in 1951, when it was isolated from skin lesions in swimmers in a contaminated swimming pool in the city of Orebro, Sweden [2]. The term "swimming pool granuloma" was coined to denote these lesions and the causal agent was classified as *Mycobacterium balnei* [3], and then, when the two mycobacteria were later seen to be identical, as *Mycobacterium marinum* (Fig. 11.1).

This mycobacterium belongs to the first of 4 groups of atypical mycobacteria in Runyon's classification (Table 11.1) [4]. It is photochromogenic (cultures turn yellowish-orange after photoexposure) (Fig. 11.2) and in culture medium (Löwenstein-Jensen) it grows slowly over 3-4 weeks, at 31°-32° C rather than 37° C. Thus, unless it is suspected and searched for

Table 11.1. Classification of mycobacteria

A. "Typical" mycobacteria	
M. tuberculosis	
M. bovis	
M. leprae	
B. "Atypical" mycobacteria	
Group I (photochromogens)	Group III (non chromogens)
M. kansasii	M. avium
M. marinum	M. intracellulare
M. ulcerans	M. xenopi
	M. triviale
	M. terrae
Group II (scotochromogens)	
M. scrofulaceum	
M. flavescens	Group IV (fast growing)
	M. fortuitum
	M. smegmatis
	M. chelonei

Fig. 11.2. Yellow photochromogenic colonies of *Mycobacterium marinum* in Löwenstein-Jensen culture medium

in appropriate culture medium, the bacillus may not be identified [5]. *Mycobacterium marinum* is alcohol- and acid-fast.

After the first observations, hundreds of other cases of human skin infections due to *Mycobacterium marinum* have been reported [6, 7]. In most cases, the infection is acquired from contaminated aquatic environments, above all swimming pools but also beaches, rivers, lakes and natural pools. Many seawater and freshwater fish have been recognized as vectors.

In 1962, the first two cases of skin infections from *Mycobacterium marinum* in tropical fish tanks were described [8]. For this reason, the affliction is also known as "fish tank granuloma". In most cases the infection onsets after cleaning the tank, although it can develop even after immersion of an arm, or after inadvertent abrasions from sharp points on the tank or from contaminated fish.

The infection has occasionally been reported in professional environments (fishermen, laboratory technicians, workers in charge of maintenance of fish tanks at aquariums, zoos or pet shops) and one case was observed in an industrial plumbing mechanic, without direct aquatic exposure [9]. Despite the various possible sources of contagion, this mycobacteriosis seems to originate most often in swimming pools. Chlorinated water does not seem to offer sufficient protection, since the micro-organism seems to be resistant to chlorine: to inhibit its growth, concentrations of 10 mg/l of water or more are required [10].

Mycobacterium marinum invades the tissues through pre-existing broken skin. The sites most commonly affected are the backs of the hands in fish tank granulomas, and the knees, elbows, arches of the feet and backs of the hands in swimming pool granulomas. The incubation period ranges from 2 weeks to 3 months. The initial lesion, generally single, presents as a reddish or reddish-blue nodule, of a soft consistency and variable diameter that may even be as large as 5-6 cm. Ulceration or colliquation may develop and the lesion will then rupture and exude pus, or else it may remain as a verrucous surface lesion. Multiple or disseminated lesions on the trunk or limbs are rare

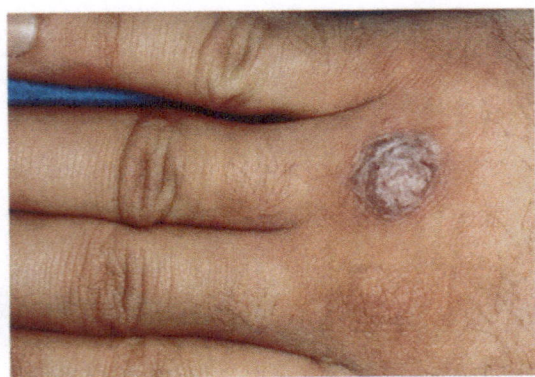

Fig. 11.3. Fish tank granuloma. Reproduced with permission from [18]

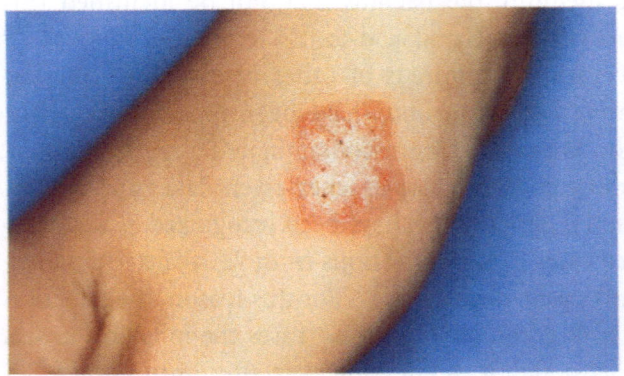

Fig. 11.4. Fish tank granuloma. Reproduced with permission from [18]

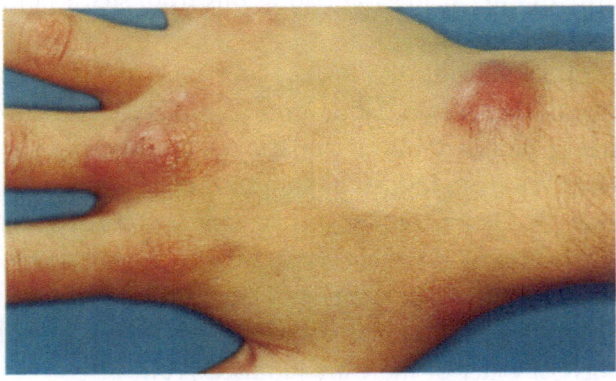

Fig. 11.5. Sporotrichoid fish tank granuloma

[11-13], except in subjects with immune deficiency. Sporotrichoid forms are frequently observed, with several nodules running along the lymphatic drainage lines [14]. One case of sporotrichoid infection of the face has also been reported in a 2-year old child, probably caught from fish in an aquarium [15].

Of the 10 cases we have observed in subjects in charge of maintenance of fish tanks, 2 had a single lesion of the hand (Figs. 11.3, 11.4) and 8 a sporotrichoid picture involving the hand and forearm (Figs. 11.5-11.7). In 3 of the latter cases there was an ulcerative evolution of the older nodular

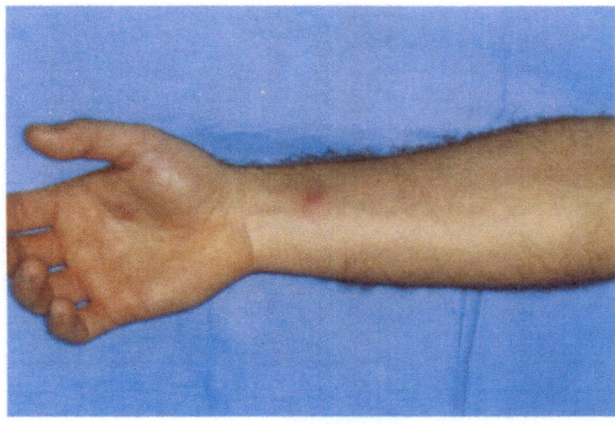

Fig. 11.6. Sporotrichoid fish tank granuloma

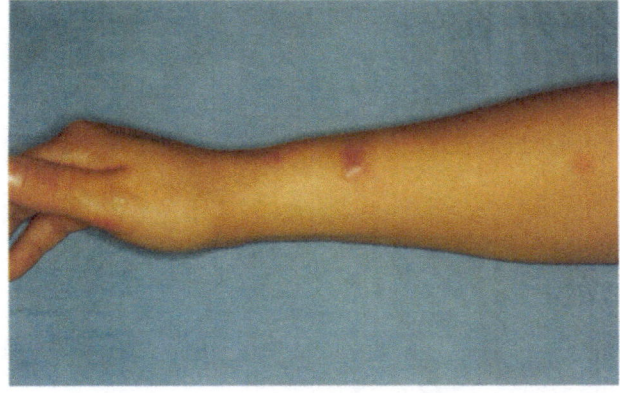

Fig. 11.7. Sporotrichoid fish tank granuloma

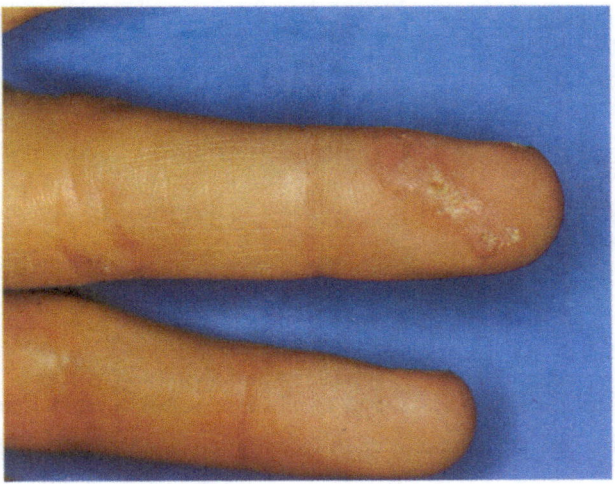

Fig. 11.8. Fish tank wound. Reproduced with permission from [18]

lesions [16-18] (Figs. 11.8-11.11). Mild involvement of regional lymph nodes is a possibility. The infection may resolve spontaneously within a few months but can persist for many years.

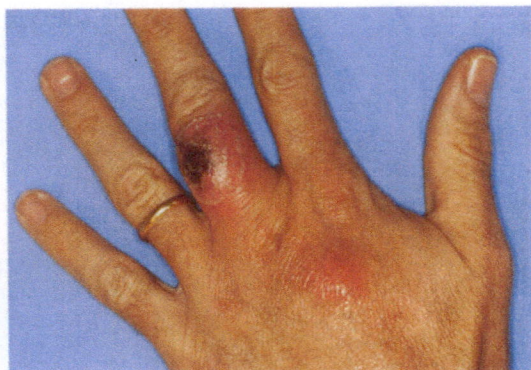

Fig. 11.9. The same case as in Fig. 11.8. Sporotrichoid fish tank granulomas, one with an ulcerative evolution. Reproduced with permission from [18]

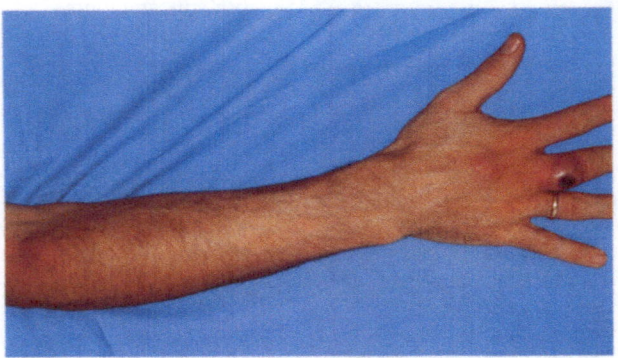

Fig. 11.10. The same case as in Fig. 11.9

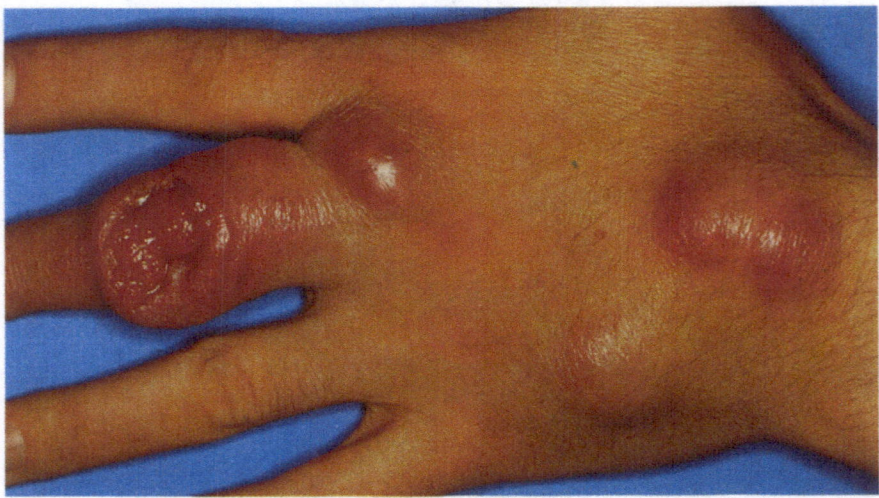

Fig. 11.11. Sporotrichoid fish tank granulomas with an ulcerative evolution. Reproduced with permission from [18]

Some interesting data emerged from a retrospective study of 38 skin infections from *Mycobacterium marinum* carried out over a period of 3 years [19]. Of the 38 patients, 30 were men and 8 women; age ranged from 14 to 85 years with a mean age of 44.7 years. The disease lasted between 1 and 132 months (mean 19 months). Thirteen subjects (34.2%) kept fish as a hobby and 4 (10.5%) had professional exposure to fish. Twelve patients (31.5%) had a history of trauma at the site of onset of the disease. In 5 (13.2%) of the 38 cases, alcohol-acid-fast bacilli were identified and in 1 (2.9%) of 35 cases, mycobacteria were isolated. Nineteen patients had been treated with cotrimoxazole-trimethoprim, 3 with minocycline, 5 with minocycline and cotrimoxazole-trimethoprim, 7 with various drug combinations and 1 with surgical excision. Three subjects had followed no therapeutic scheme. Clinical improvement was observed in 26 subjects (68.4%), 2 (5.3%) had no response and 10 (26.3%) were lost to follow-up. None of the patients suffered worsening of the symptoms during treatment [19].

Mycobacterium marinum, antigenically correlated to *Mycobacterium tuberculosis*, does not confer immunity and re-infection is possible. The clinical history is very important for diagnosis, although even tuberculosis can be contracted at a swimming pool. The infection must be differentiated from leishmaniasis, sporotrichosis, primary cutaneous tuberculosis and papulo-verrucous tuberculosis.

Identification of the mycobacterium in culture is a must. However, the culture is positive in only about 70% of cases. In the near future, polymerase chain reactions will be routinely used to ensure rapid, sensitive and specific diagnosis. Intradermal tests with PPD prepared from *Mycobacterium marinum* are positive. The histological picture is characterized by a tuberculoid granuloma, with evident fibrinoid rather than caseous masses; giant cells are not always a feature. Sometimes the findings are less specific. In histological preparations, alcohol-acid-resistant bacilli can be identified using Ziehl-Neelsen or trichromic staining.

Resistance to chemotherapy is common and an antibiogram is useful to guide treatment. The drugs found most effective are minocycline, sulphamethoxazole, tetracycline, rifampicin, isoniazide and claritromycin [15].

Personal experience

As mentioned above, we have observed 10 cases of granulomas induced by *Mycobacterium marinum* [16-18]. In all the subjects the affliction followed trauma during maintenance of fish tanks. In one case with a sporotrichoid picture, the dermatitis onset 3 months after a wound from a sharp stone on the bottom of the tank at the level of the tip of the third finger of the left hand (Fig. 11.8). In all cases the incubation time ranged from 2 to 4 months.

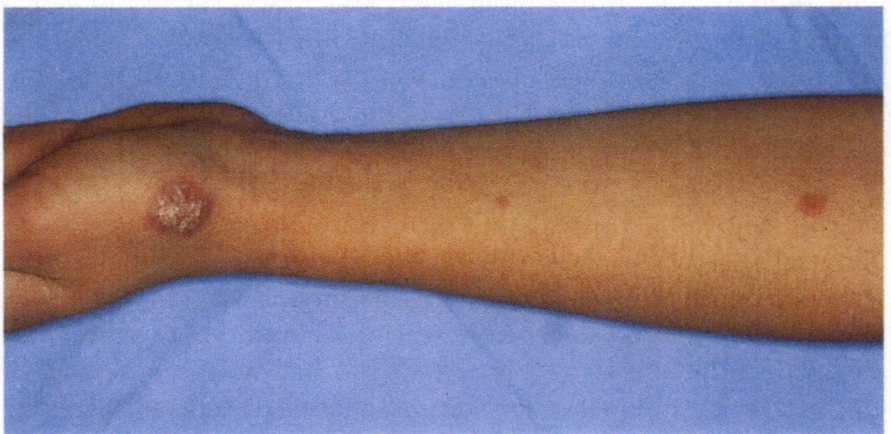

Fig. 11.12. Fish tank granuloma with intensely positive PPD to *Mycobacterium marinum* and weakly positive PPD to *Mycobacterium tuberculosis*

Intradermal tests with PPD from *Mycobacterium marinum* were strongly positive in all 10 subjects (Fig. 11.12), while tests with PPD from *Mycobacterium tuberculosis* elicited only a weak reaction.

Cultures of biopsies or aspirated material from the sites of the lesions in Löwenstein-Jensen culture medium at 30° C gave rise to the growth of yellowish-orange photochromogenic colonies after 15 days in only 8 cases. In the literature, negative responses to culture tests have been observed in about 30% of cases. In one case we were also able to isolate the organism in dead fish in the tank and in another, in the fish tank filter. To identify the bacterial genus, the colonies grown in Löwenstein-Jensen culture medium were subjected to a series of tests according to the methods proposed by the Atlanta CDC; the results of these tests are reported in Table 11.2.

As to treatment, 1 case with a sporotrichoid picture did not respond to sulphamethoxazole-trimethoprim but resolved after 2 months of treatment with rifampicin (900 mg per day) and isoniazide (600 mg per day). In the other 9 cases the infection regressed after treatment with minocycline (200 mg per day) for 2-4 months.

Infection from other mycobacteria

A new species of mycobacterium has recently been isolated from skin ulcers and granulomas on various internal organs of bass in Chesapeake Bay, Maryland (USA). The bacterium is slow-growing at 28° C, is not chromogenic, does not show any nitrate-reducing activity, catalase activity, hydrolysis in Tween 80 or arysulphatase-reduction. It grows best at low saline

Table 11.2. Differential features of the mycobacteria inducing skin lesions

Species	M. ulcerans	M. tuberculosis	M. marinum	M. flavescens	M. szulgai	M. fortuitum	M. chelonei
Optimal growth temperature	32° C	37° C	25-30° C	37° C	37° C	37° C	
Pigmentation	N	N	F	S	S-F	N	N
Niacin test	–	+	V	–	–	–	V
Reduction of nitrates	–	+	–	+	+	+	–
Tween 80 hydrolysis (5 days)	–	–	+	+	V	V	V
Semiquantitative catalase (> 45 mm)	–	–	–	+	+	+	+
Catalase 68° C pH 7	+	–	–	+	+	+	V
Urease	–	+	+	+	+	+	+
Pirazinamidase	–	+	+	+	+	+	+
Growth 5% NaCl	–	–	–	+	–	+	V
Growth Mac Conkey Agar	–	–	–	–	–	+	+

N, non photochromogenic; *S*, scotochromogenic; *F*, photochromogenic; *V*, variable

concentrations and is positive to urease and pyrizinamidase. PCR demonstrated a single nucleotide sequence, equal to no other sequence recorded in any database. Analysis of the almost complete sequence that codes for rRNA 16 S showed a single nucleotide repetition, with an 87.7% homologous structure with *Mycobacterium ulcerans*, 87.6% with *Mycobacterium tuberculosis* and 85.9% with *Mycobacterium marinum*. Phylogenetic analysis located the bacterium within the *Mycobacterium tuberculosis* complex [20].

Erysipeloid

This dermatitis, also known by the name of Baker-Rosenbach's erysipeloid, is an acute, rarely chronic infection induced by *Erysipelothrix rhusiopathiae*, the aetiological agent of "swine rotlauf disease". *Erysipelothrix rhusiopathiae* is a Gram-positive, non spore-producing and non mobile bacillus, which usually has long filaments. It can survive in the environment for long periods and also lives in the sea. The infection is common not only in pigs but also in horses, chickens, ducks, sheep, turkeys and other animals and in salt and freshwater fish.

Although it is widespread all over the world in animals, man rarely contracts it. Most of the cases reported were due to professional activities, being most often observed in fishermen and butchers. It has also been described in housewives pricked by fish or chicken bones.

Erysipeloid generally onsets in late summer when animal infections are most common. About 3 days after contagion, the puncture zone develops a dark, erythematous raised area with an irregular centrifugal

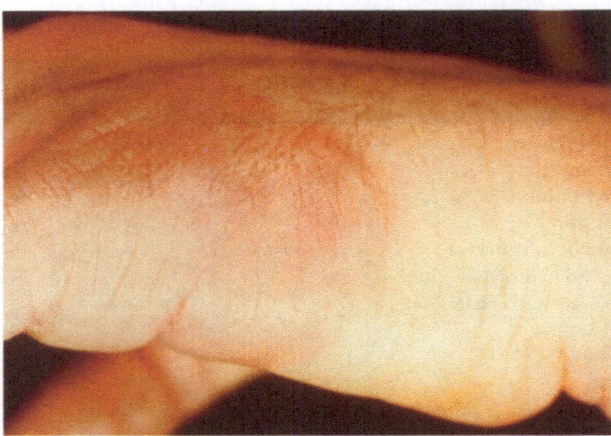

Fig. 11.13. Erysipeloid. Reproduced with permission from [18]

extension and distinct, raised polycyclical margins (Fig. 11.13). The sites most commonly involved are the hands and forearms but all exposed areas may be affected. In 10% of cases, fever onsets; pricking and itching sensations and pain may be present. The area affected will spread wider over the following days and reach a maximum diameter of 10 cm. It resolves spontaneously in 2-3 weeks without any desquamation or suppuration phenomena [21-24].

Apart from this modest, localized form, a generalized skin condition can be observed in rare cases, as well as systemic forms complicated by septicaemia and endocarditis. Differential diagnosis must be made with erysipelas, a febrile streptococcal infection that spreads rapidly. The disease does not confer immunity and re-infections are therefore possible. The bacterium responsible can be cultured from a biopsy sample obtained from the margins of the skin lesions or from peripheral blood in systemic forms. However, isolation and identification is still not easy as the organism lies deep in the skin and culture times are rather long. Some progress has recently been made in molecular approaches for diagnosis and for taxonomic and pathogenic studies of *Erysipelothrix*. Two different PCR techniques have also been described for diagnosis of the swine infection, one of which has since been used successfully in human samples [25]. The infection is treated by administering a week's course of penicillin or tetracycline.

Infected wounds

Sea water is a dilute suspension of bacteria, many of which are harmful to man and can cause various infections (external otitis, gastroenteritis, pneumonia). Wounds from any cause occurring in the sea can therefore easily become infected. The micro-organisms isolated from infected wounds include bacteria present in seawater (vibriones, enterobacter, *Escherichia coli, Pseudomonas, Achromobacter xylosoxidans, Acinetobacter calcoaceticus*) and bacteria from normal skin flora (staphylococci, streptococci).

The treatment of wounds infected by marine organisms includes debridement of necrotic tissue, removal of foreign bodies, drainage of abscesses and irrigation with saline or antiseptic solutions. If necessary, anti-tetanus injections or systemic antibiotic treatment should be given (tetracycline or aminoglycosides), although the efficacy of such treatments on some vibriones (*Vibrio vulnificus*) has not yet been established.

References

1. Aronson JD (1926) Spontaneous tuberculosis in salt water fish. J Infect Dis 39:315
2. Norden A, Linell F (1951) A new type of pathogenic mycobacterium. Nature 168:826
3. Linell F, Norden A (1954) *Mycobacterium balnei*: new acid-fast bacillus occurring in swimming pools and capable of producing skin lesions in humans. Acta Tuberc Scand 33:1
4. Runyon EH (1959) Anonymous mycobacteria in pulmonary disease. Med Clin North Am 43:273
5. Bhatty MA, Turner DP, Chamberlain ST (2000) *Mycobacterium marinum* hand infection: case reports and review of literature. Br J Plast Surg 53:161
6. Philpott JA, Woodburne AR, Philpott OS et al (1963) Swimming pool granuloma: a study of 290 cases. Arch Dermatol 88:158
7. Huminer D, Pitlik SD, Block C et al (1986) Aquarium-borne *Mycobacterium marinum* skin infection. Arch Dermatol 122:698
8. Swift S, Cohen H (1962) Granulomas of the skin due to *Mycobacterium balnei* after abrasions from a fish tank. N Engl J Med 267:1244
9. Cole GW (1987) *Mycobacterium marinum* infection in a mechanic. Contact Dermatitis 16:283
10. Reznikov M (1970) Atypical mycobacteria. Their classification and aetiological significance. Med J Aust 1:553
11. Gould WM, Mc Meekin DR, Bright RD (1968) *Mycobacterium marinum* (*balnei*) infection. Arch Dermatol 97:159
12. King AJ, Fairley JA, Rasmussen JE (1983) Disseminated cutaneous *Mycobacterium marinum* infection. Arch Dermatol 119:268
13. Blank AA, Schnyder VW (1985) Aquariumgranulom vom kutan-disseminierten Typ. Hautarzt 86:48
14. Alessi E, Finzi AF, Prandi G (1976) Infezioni cutanee da *Mycobacterium marinum*. Considerazioni su due casi clinici. G Ital Min Derm 111:85
15. Brady RC, Sheth A, Mayer T et al (1997) Facial sporotrichoid infection with *Mycobacterium marinum*. J Pediatr 130:324
16. Angelini G, Filotico R, De Vito D et al (1990) Infezioni cutanee professionali da *Mycobacterium marinum*. Boll Dermatol Allergol Profes 5:165
17. Angelini G, Vena GA, Grandolfo M (1995) Micobatteri e cute. Dermotime 7:11
18. Angelini G, Vena GA (1997) Dermatologia professionale e ambientale. Vol I. ISED, Brescia

19. Ang P, Rattana-Apiromyakij N, Goh CL (2000) Retrospective study of *Mycobacterium marinum* skin infections. Int J Dermatol 39:343
20. Heckert RA, Elankumaran S, Milani A et al (2001) Detection of a new mycobacterium species in wild striped bass in the Chesapeake Bay. J Clin Microbiol 39:710
21. Meneghini CL, Angelini G (1981) Dermatosi professionali. In: Sartorelli E (ed) Trattato di Medicina del Lavoro. Vol II. Piccin, Padova, 986
22. Burnett JW (1962) Uncommon bacterial infections of the skin. Arch Dermatol 86:597
23. Barnett JH, Estes SA, Wirman JA et al (1983) Erysipeloid. J Am Acad Dermatol 9:116
24. Robson JM, McDougall R, van der Valk S et al (1998) *Erysipelothrix rhusiopathiae*: an uncommon but ever present zoonosis. Pathology 30:391
25. Brooke CJ, Riley TV (1999) *Erysipelothrix rhusiopathiae*: bacteriology, epidemiology and clinical manifestations of an occupational pathogen. J Med Microbiol 48:789

12 Aquatic skin diseases from physical and chemical causes

The hydrosphere and man

A part from the diseases due to biotic noxae described up to now, there are various other dermatological conditions connected in some way with salt or freshwater contact or with aquatic activities (Table 12.1) [1, 2]. No analysis will here be made of afflictions favoured, induced or aggravated by exposure to the sun, which is in a sense obligatory in subjects involved in aquatic activities for long or short periods, or of those which may benefit from such exposure, as they are outside the scope of this work.

Table 12.1. Some skin diseases of aquatic origin

General	Seawater
Sunburn	Immersion syndrome
Aquagenic urticaria	Abrasive effect
Aquagenic pruritus	Surfer's nodules
Cold urticaria	Otitis externa
Contact dermatitis	
Swimming costume	**Sub-aqua activities**
Diving equipment	Otitis externa
	Intertrigo
Freshwater	Staphylococcal infections
Swimming pool	Burns
Mycosis	Linear abrasions from wetsuit folds
Verrucae	Pruritus and erythema from
Chlorine irritation	decompression
Chapping in atopic subjects	"Napkin rash" type dermatitis
Greenish hair tinge	
Hair bleaching	
Chemical conjunctivitis	
Otitis externa	
Jacuzzi/hot tubs	
Folliculitis induced by *Pseudomonas*	
Sauna	
Miliaria	
Tinea versicolor	
Shower	
Aquagenic pruritus	

Aquagenic urticaria

This form of urticaria is induced simply by skin contact with the water, regardless of its physical and chemical properties (source, salt content, temperature) [3-10]. The affliction is often misdiagnosed. It generally onsets in young adults, with a mean age of 18 years, and is five times as frequent in the female sex. Sometimes, several members of the same family are affected. There are no data available on the evolution and duration of the disease, although many patients have referred a very long course, even 20 years.

The urticarial infection onsets after 3-10 minutes from any type of skin contact with water; it reaches a peak in about 30 minutes and dies down again after a further 30-60 minutes. The exposed skin sites are generally refractory to stimuli for several hours. The raised lesions are not clinically distinguishable as regards shape and distribution from those of cholinergic urticaria (and can coexist with the latter), being punctiform, 2-3 mm in diameter, surrounding the hair follicles. These small wheals, that are intensely itchy, appear on the areas in contact with the water and if the whole body is exposed, mainly on the neck and trunk, and to a lesser degree the shoulders and sides. The palms of the hands and soles of the feet are not affected. No systemic reactions or alterations in laboratory parameters have been reported.

The pathogenetic mechanism is unknown. Two mechanisms seem most likely to be implicated: the first is a cholinergic reaction consisting of local release of acetylcholine. Pre-treatment of the exposed skin sites with an acetylcholine antagonist, scopolamine, has been shown to suppress the urticarial response. The other mechanism is mast-cell degranulation. In fact, during urticarial crises, increased blood levels of histamine and the presence of degranulated mast-cells have been demonstrated in the affected skin. Degranulation could be triggered by the action of a molecule deriving from interaction between the water and some substances present in the skin, probably of a sebaceous or sudoriparous nature [11]. Czarnetzki and coll. have hypothesized that a water-soluble antigen of epidermal origin may spread in the derma on contact with water (which would not therefore be the causal agent but merely a vehicle), giving rise to the release of histamine by the sensitised mast-cells, presumably due to the presence of specific IgE on the membrane [6].

The easiest diagnostic method is bathing in water at 35°-36° C for 30-40 minutes. Alternatively, compresses soaked in tap or distilled water can be applied to the patient's chest for 40 minutes. It is best to use a thermometer on the site of the test to make sure of the correct temperature, in order to be able to differentiate between aquagenic, cholinergic, cold and heat urticaria.

Table 12.2. Skin reactions directly (*) or indirectly (°) (pseudo-aquagenic reactions) due to contact with water

Reaction	Stimulus
Aquagenic pruritus (*)	Water
Aquagenic urticaria (*)	Water
Cholinergic urticaria (°)	Bathing or swimming in hot water
Acquired cold urticaria (°)	Cold water
Localized heat urticaria (°)	Hot water
Symptomatic dermographism (°)	Shower jet, towel friction
Polycythaemia rubra vera (*)	Water

A possible association with dermographism should be taken into account. If only itching develops, without skin lesions, then a diagnosis of aquagenic pruritus can be made.

There must also be differential diagnosis with other complaints induced by water, although not always directly (Table 12.2). Apart from aquagenic urticaria, other cases of urticaria from physical causes develop only after contact with certain types of water, which is not the true cause. Pre-treatment of the skin with vaseline or lanoline prevents the eruption, perhaps by preventing formation of the causal molecule.

The treatment of choice is antihistamines, although variable responses are obtained. If taken before contact with water, they can prevent or relieve the symptoms in many cases. Sometimes, the refractory period may be exploited to induce tolerance (as in other forms of physical urticaria), after provoking lesions. PUVA-treatment has been employed in association with astemizole [12].

Cold urticaria

This consists of the appearance of pomphoid lesions in skin or mucosal sites after contact with cold objects, water, or air, or after ingestion of cold drinks or foods. Among the forms of physical urticaria, cold urticaria can present under various guises: familial or acquired, immediate or delayed, localized or systemic, primitive or secondary (Table 12.3). Its incidence ranges from 1% to 7% of all physical forms, depending on the various case series [7-9].

Table 12.3. The various forms of cold urticaria [9]

Familial
Localized (delayed, after 9-18 hours)
Systemic (immediate)

Acquired
Localized
Immediate
Delayed
Cholinergic cold dermographism
Due to localized reflex
Systemic
Due to cold
Due to generalized reflex
Cholinergic cold urticaria
Idiopathic
Secondary
Viral infections (respiratory virosis, infectious mononucleosis)
Insect sting
Drug intolerance or allergy (penicillin, griseofulvin)
Atopy
Diseases featuring cryoglobulins, cryofibrinogen or haemolysins due to the cold
Autoimmune diseases (connectivitis, thyroiditis, erythema nodosum)
Cold paroxystic haemoglobinuria
Myeloproliferative diseases (myeloma)

The acquired idiopathic form is the most common and the familial form the rarest; delayed and cold cholinergic forms (in which puncti-form pomphoid lesions 1-2 mm wide develop after physical exercise in a cold environment, whereas no clinical lesions appear after physical exercise in a warm environment or else exposure to a cold environment but no physical exercise) are also exceptional. Another rare form is cholinergic cold dermographism (linear pomphoid lesions onset if the patient is exposed to the cold during or immediately after applying appropriate stimulation; exposure to the cold without such stimulation or alternatively the application of stimulation in a warm environment, do not elicit any response).

Acquired cold urticaria may be secondary to various conditions: viral infections (mononucleosis), reactions to insects, intolerance or allergy to drugs (penicillin, griseofulvin), diseases with cryoglobulins, cryofibrinogen or cold haemolysins (cold agglutinins are not associated with urticaria).

Acquired idiopathic cold urticaria can onset at any age but it particularly affects young adults, especially the female sex. The classic picture is the development of pomphoid lesions in sites exposed to contact with cold objects, foods, air conditioning or sudden changes of temperature. Uncovered skin zones seem to be much more sensitive to stimuli. The skin wheals usually develop within a few minutes and persist for 1-2 hours. The oral mucosa and tongue may also be affected. In more severe cases with diffuse manifestations, systemic symptoms such as weakness, breathlessness, headache, tachycardia and vertigo can develop. Sometimes even shock symptoms can onset immediately after swimming and it is very important to warn patients suffering from cold urticaria of the dangers of swimming.

Pathogenetic considerations should bear in mind that cold urticaria can also be passively transferred to healthy subjects by means of the Prausnitz-Küstner test; this passive sensitisation has to do with serum IgE, and sometimes IgM, IgG or IgA [13]. It has been suggested that subjects with cold urticaria develop auto-antibodies of IgE type, but also IgG, against skin antigens prevalently associated with the skin mast-cells. Anti-nuclear serum auto-antibodies (acting against the b laminar fraction) have occasionally been demonstrated [14]. The efficacy of antibiotic treatment demonstrated in a high percentage of patients has recently given rise to the suggestion that acquired forms may have a microbial origin, even of a subclinical nature [15].

Histamine, a chemotactic neutrophilic factor and some chemotactic eosinophilic factors are among the main mediators of cold urticaria. In the acquired idiopathic form, tests of immersion in cold water have instead elicited a decreased chemotactic neutrophilic index; this event, described as "the granulocytic inactivation phenomenon" seems to be highly specific and is not present either in chronic urticaria or in other forms of cold urticaria [8]. Quinines and some derivatives of arachidonic acid (PGD2, LTE4) sometimes appear to be increased; they may amplify the biological effect of other mediators [16, 17].

The observation that topical capsaicin, an antagonist of substance P released at the level of the nerve terminals, can inhibit local reactions to contact with the cold implies that this neuropeptide may play an important role in triggering the complaint. On the basis of these data, defective thermal and/or vasomotor regulation of central origin may be supposed [8]. During severe episodes of cold urticaria, associated with systemic symptoms, high serum levels of TNF-α, a powerful pro-inflammatory cytokine, have been demonstrated [18].

At histology, the inflammatory dermal infiltrate features two different cellular patterns, one with a predominance of neutrophils and the other of lymphocytes. This is likely to be the result of two different stages of the same process: the neutrophils may prevail at first, and then the lymphocytes take over [19].

Cold urticaria can be associated with chronic urticaria or other urticarial forms with physical causes, especially dermographism and cholinergic urticaria. The diagnosis is often confirmed by the ice-cube test: when applied to the flexural surface of the forearm for 5-20 minutes, an itchy skin wheal will develop on the site of the test, of exactly the same size and shape as the ice cube. This test is positive in 90% of cases; if it is negative, other procedures can be tried, such as applying a cooled metal cylinder, or immersing the forearm for 5-15 minutes in cold water (0°- 8° C) or exposure to a cold room (+ 4° C). The ice-cube test is negative in familial and cholinergic cold urticaria forms.

The clinical course of acquired idiopathic cold urticaria is quite long, ranging from 2 to 10 years.

As to treatment approaches, the patient must first of all be carefully informed about all the possible triggering factors and especially the danger of swimming in cold water (in any case, such patients must never swim unaccompanied). Patients must also protect themselves against cold air with appropriate clothing, including gloves and woollen socks. It may be possible to induce tolerance to the cold, by repeated immersion in water at 10°-15° C for 5 minutes at intervals of 1 hour at first, and then every 2, 4, 6, 8, 12 and 24 hours.

Among antihistamines, cyproheptadine is particularly effective. Loratadine, cinnarizine, cetyrizine and chetotyphene (2-4 mg/die) are also helpful. Doxepine (10-25 mg, 3 times a day), a tricyclical anti-depressant, has been found to work better than hydroxyzine, cinnarizine and cyproheptadine [20]. Systemic corticosteroids are not successful. Antibiotic treatment (penicillin and tetracycline) has yielded good results in about two thirds of cases [15].

Familial cold urticaria has dominant autosomal transmission. Pomphoid lesions appear, associated with burning rather than itching, between 30 minutes and 3 hours after exposure to cold wind but not contact with cold objects: the ice-cube test is negative. General symptoms often develop: shivering, fever, muscle and joint pain, headache. The symptoms are present from birth and persist throughout life. The pathogenetic mechanism is not clear, and the passive transmission test is negative. Leukocytosis is generally present and skin biopsy reveals a polymorphonuclear infiltrate. Diagnosis is on the basis of a positive family history, onset at birth, a negative ice-cube test and the presence of systemic signs. In some cases, stanazole, an attenuated androgen, can help [21].

Aquagenic pruritus

Showering or bathing are very enjoyable daily interludes, essential for hygiene and sometimes as a form of treatment. However, some subjects suffer acute pruritus as a direct consequence of bathing. In some of these, contact with water is an indirect stimulus (acute pseudo-aquagenic reactions) (Table 12.2), while in others the pruritus is a direct local effect of skin contact with water, as in aquagenic urticaria (featuring objective signs), and aquagenic pruritus (no objective signs).

The latter, in turn, may be:

1. true aquagenic pruritus (Table 12.4);
2. senile aquagenic pruritus (Table 12.5);
3. aquagenic pruritus observed in 50% of patients with polycythaemia rubra vera (Table 12.6).

True aquagenic pruritus features intense pricking or burning pruritus that develops after contact with water, regardless of temperature, with no apparent objective signs. It lasts between 10 and 120 minutes. In some subjects it manifests during bathing, while in others it onsets immediately after coming out of the water. In almost all cases the legs and especially the thighs are mainly affected, although the trunk and arms may also be involved [7, 22, 23].

Although the release of histamine and mast-cell degranulation have been demonstrated, histamine does not seem to be the main mediator. Intradermal injections of acetylcholine do not reproduce the itching symptoms. An increased fibrinolytic cutaneous activity has recently been shown, although the plasma fibrinolytic activity was within normal limits [24]. This increased fibrinolytic cutaneous activity can be blocked by ε-aminocaproic acid, suggesting that the increased activity may be due to an inherently increased plasminogen activity. Intradermal histamine and acetylcholine injections trigger increased cutaneous fibrinolytic activity, which implies that the phenomenon may be secondary to histamine and acetylcholine release.

Table 12.4. Diagnostic criteria for aquagenic pruritus

1. Intense, recurrent itching after contact with water, regardless of temperature
2. Itching onsets within a few minutes after contact with water and may persist for up to 2 hours
3. No visible skin signs
4. Cold, aquagenic, cholinergic heat and cold, localized heat urticaria forms and symptomatic dermographism have been excluded
5. No skin or internal disease is present, no likely drugs are being taken
6. Polycythaemia rubra vera has been excluded

Table 12.5. Diagnostic criteria for senile aquagenic pruritus

1. Especially in the female sex (75%) in subjects over 60 with pale skin
2. Excessively dry skin
3. Intense itching after drying
4. The intensity of the itching sensations is proportional to the duration of exposure to water and the degree of skin dryness
5. Triggering factors: contact with water and consequent dryness of the skin, variations in temperature, friction
6. The itching starts on the legs or forearms, then spreads and persists for 10-60 minutes
7. The intensity of the itching increases with age and during the winter

Table 12.6. Diagnostic criteria for aquagenic pruritus associated with polycythaemia rubra vera

1. Only subjective signs as in aquagenic pruritus
2. Specific symptoms of the disease
3. More frequent onset of spontaneous itching
4. Itching generally depends on the temperature of the water
5. Hot baths elicit worse itching than cool baths
6. Cooling of the skin provokes itching
7. The intensity of the itching is not correlated with the severity of the disease

Diagnosis is relatively easy (Table 12.4) and can be confirmed by the provocation test. Differential diagnosis is with senile aquagenic pruritus (Table 12.5) and aquagenic pruritus observed in 50% of subjects with polycythaemia rubra vera (Table 12.6).

Aquagenic pruritus does not respond well to antihistamines, that only partially relieve the symptoms. About 50% of patients have at least a temporary response to sub-erythematogenic doses of UVB rays (290-320 nm) 3 times a week. Emollients and sodium bicarbonate added to the water reduce the itching symptoms in some cases; the application of vaseline to the skin before bathing can also help. However, the best results are obtained simply by limiting contact with water as much as possible (Table 12.7).

Table 12.7. Treatment of water-induced pruritus

AP	SAP	PVAP
Softeners	Softeners	Specific treatment
Limiting contact with water	Limiting contact with water	Aspirin
UVB rays	Avoiding soap	Antihistamines
Sodium bicarbonate	Moisturisers	

AP, aquagenic pruritus; *SAP*, senile aquagenic pruritus; *PVAP*, polycythaemia vera with aquagenic pruritus

Contact dermatitis

Contact dermatitis caused by the swimming costume is rare but possible, due to sensitisation to the elastic (rubber additives) or dyes.

Contact dermatitis from diving equipment can be observed, due to professional or sports activities, and induced by the mask (Figs. 12.1, 12.2), goggles, snorkel, fins and rubber wetsuits [25-27]. The sensitising substances responsible are mercaptobenzothiazole, tetramethylthiouram disulphide, thiourams, dithiocarbamates, paraphenylenediamine, thiourea, formaldehyde, butylphenolformaldehyde resin, isopropyl-phenyl-paraphenylenediamine.

In sensitised subjects, snorkels can also cause inflammation of the mouth, that generally starts with an intermittent, mild burning sensation associated with ingesting hot drinks or spicy foods.

Saltwater dermatitis

Prolonged immersion in seawater causes electrolytic alterations due to percutaneous absorption (immersion syndrome). Occasionally, a skin peeling effect may appear where the swimming costume hugs closely, that may even evolve into ulceration due to the combination of friction and the abrasive effect of the salt.

Surfer's nodules are hard and indolent and onset at the level of the anterior tibial region. They are reversible and caused by continual contact with the board. These pretibial fibrotic masses may be dermal or hypodermal and can also result in deformity of the underlying bone and calcifications [28].

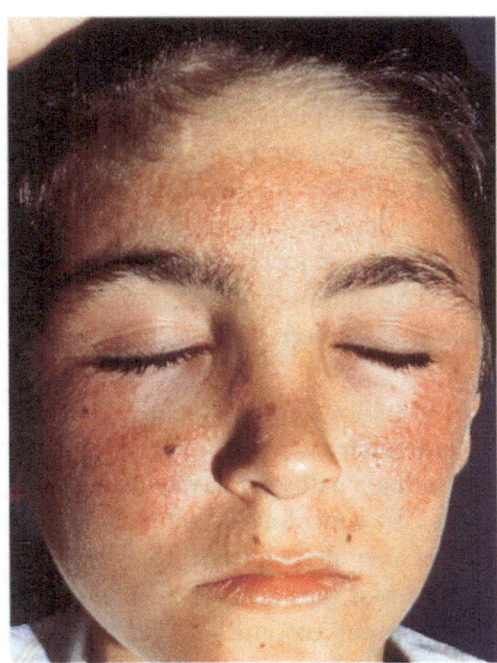

Fig. 12.1. Allergic contact dermatitis to rubber mask. Reproduced with permission from [27]

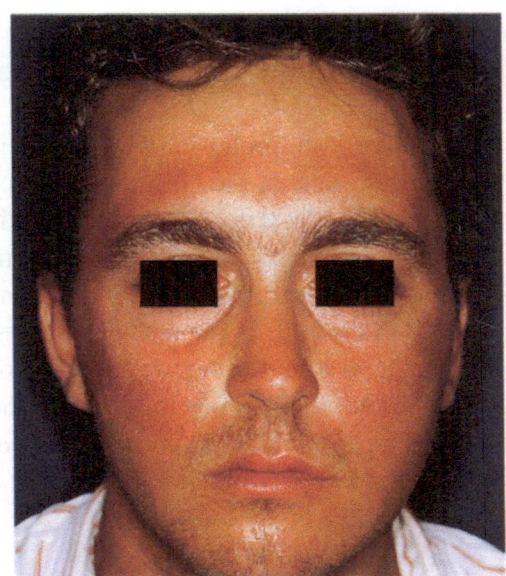

Fig. 12.2. Allergic contact dermatitis to rubber mask. Reproduced with permission from [27]

External otitis is an acute bacterial infection of the external ear fostered by the macerating effect of the water and the persistent humidity of this part. The clinical symptoms are pain, exudation, pruritus and sometimes fever and impaired hearing. The most common bacterial cause is *Pseudomonas aeruginosa* [29].

Freshwater dermatitis

Swimming in chlorinated pools has a dehydrating effect on the skin and hair (anti-oil action) that is more evident in atopic subjects. Depending on the concentration, chlorine has a bleaching effect on the hair, which is most apparent in blonde subjects and in the summer months in combination with the sunrays. A greenish tinge may develop in blonde subjects who often swim in strongly chlorinated pools; shampooing the hair immediately after swimming is the best prophylaxis. Temporary chemical conjunctivitis (so-called "red eyes") is observed in subjects who swim with their eyes open.

Skin eruptions have been reported in subjects using a Jacuzzi. These are forms of folliculitis that onset 8-48 hours after exposure. The aetiological agent isolated in high concentrations in these baths is *Pseudomonas aeruginosa*. The high temperature of the water is a contributing factor as it causes the pores to dilate. The eruption resolves after about a week. The lesions are of erythemato-papulous and vesico-papulous type and are itchy. The sites affected are mainly those under the swimming costume and the head and neck are never involved, being out of the water.

Dermatitis associated with deep sea diving

Notoriously, both professional and amateur scuba divers are exposed to an enormous variety of risks including skin problems [30].

Staphylococcal skin infections are relatively frequent and also difficult to treat. Overheating inside the wetsuit can cause local burns. Underwater welding procedures can induce erythema and telangiectasia. The skin trapped in the folds of the wetsuit can present linear abrasions.

During decompression, divers may notice itching, with or without an urticarial eruption, mainly localized on the back or trunk [31]. If they stay underwater for long they may develop a form of "napkin rash", due to having to attend to physiological needs.

Cutaneo-systemic complaints in fishermen

Deep-sea fishermen can be victims of rare, practically unthinkable events such as a possible encounter with a bomb containing mustard gas. From 1970 till now, we have observed 11 fishermen with dermatitis that onset from 6 to 10 hours after fishing in the open sea outside Molfetta, a city 30 km to the North of Bari [32-33]. All these patients presented an intensely

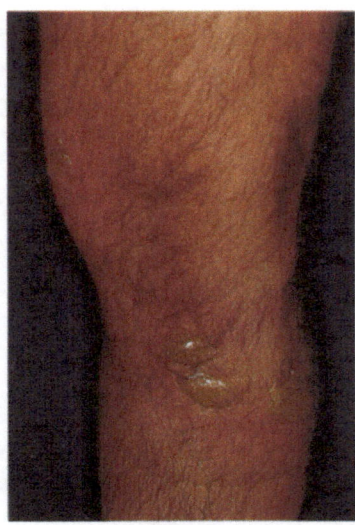

Fig. 12.3. Blistering dermatitis from mustard gas

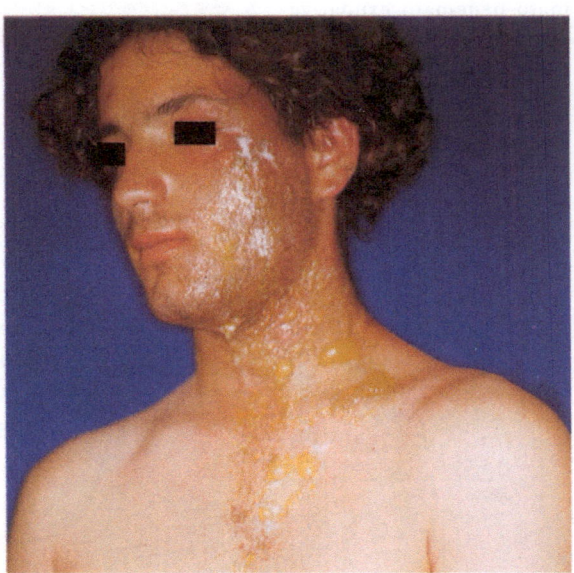

Fig. 12.4. Blistering dermatitis
from mustard gas

erythematous-oedematous dermatitis featuring widespread blistering
lesions with a clear liquid content (Figs. 12.3, 12.4). The affliction particularly involved the hands, forearms and face (where the erythema and oedema were more marked, being especially severe on the eye-lids), while in 3
cases the genitals were also affected (Fig. 12.5), with intense erythema in 2
and erythema, oedema and blisters in 1. All cases were associated with
severe conjunctivitis, lachrymation and photophobia. There was intense
burning and itching at the affected sites. In 6 patients the skin symptoms
were associated with headache, vomiting and nausea.

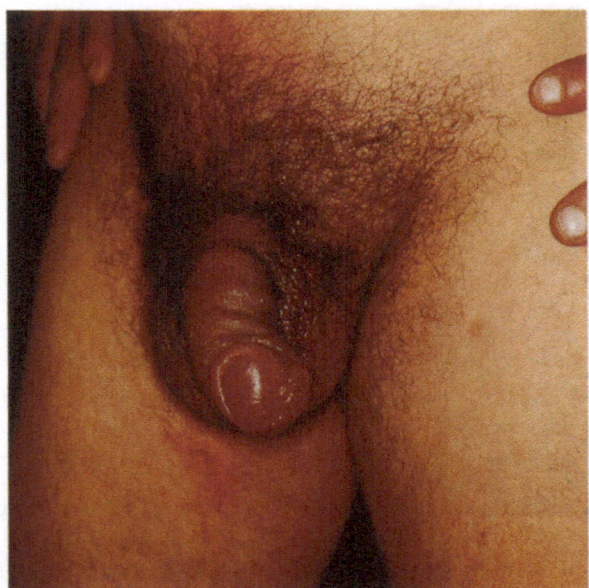

Fig. 12.5. Intensely erythe-mato-oedemato-exudative dermatitis from mustard gas

Fig. 12.6. Bombs containing mustard gas can be pulled in together with the fish in dragnet fishing

The fishermen referred that when they pulled their nets on board they had found bombs in with the fish (Fig. 12.6). A few hours after handling the nets contaminated with the liquid gas contained in the bombs, they suffered the above symptoms. The dermatitis of the hands and forearms was obviously induced by direct irritant contact with the contaminated bombs and nets, while the other skin and mucosal lesions were caused by evaporation of the gas and hence airborne cutaneo-mucosal dermatitis (irritant airborne contact dermatitis) [34]. In all the cases, the symptoms resolved after 10-15 days leaving dark skin patches. The conjunctivitis was treated with eye-baths containing 2% sodium bicarbonate and antibiotic eye-drops. The other symptoms regressed rapidly with symptomatic treatment. Controls after 20-30 days excluded any re-presentation of the dermatitis.

The risk of fishing bombs is well known to fishermen and the harbour authorities in the area. These authorities report that more than 100 cases of intoxication from mustard gas have been observed over the years. This gas (2,2'-dichlorodiethyl sulfide: $C_4H_8CI_2S$), one of the most aggressive gases used in chemical warfare, is also known as "iprite" after the city of Ypres (Belgium), where it was first used in July 1917. The English and Americans call it mustard gas because of its characteristic odour. In the pure state it is an odourless, colourless oily liquid, and the characteristic yellowish-brown colour and mustard-like smell are due to impurities (ethylsulphides). It is not very soluble in water but dissolves rapidly in organic solvents or fats; this facilitates penetration of the cells, where it has a toxic effect. It evaporates slowly because of its low vapour pressure, although this increases with higher temperatures. It is toxic both in liquid and vapour form: in the former cases it damages the skin and in the latter, the skin, conjunctiva and respiratory mucosa. Its toxic effects manifest after 4-24 hours from exposure [35, 36].

Since the First World War, intoxication from mustard gas has only been caused by occupational contact, apart from its widespread use during the Iran-Iraq war (1980-1988) [37]. The cases we observed were due to the fact that there used to be factories for loading and unloading mustard gas bombs in Molfetta. After the Second World War, the bombs were thrown into the sea about 3 miles from the coast. The bombs are therefore sometimes fished up, especially in the summer season when drag-nets are used.

Although chemical bombs are present in all European seas, similar cases of dermatitis from mustard gas have only occasionally been reported [38-40], probably because it is practically impossible to connect the disease with contamination by fishing nets unless the bombs are actually seen in them. Otherwise, the skin symptoms may be attributed to the harmful action of some marine flora and fauna. Moreover, the fishermen in our area do not always report such symptoms, sometimes because they are only mild (and in any case they know about the problem) and sometimes for fear of their catch being confiscated or because they intend to keep the bombs for illegal explosive fishing.

Fishermen should be informed of the risks of fishing bombs in particular areas, and must be instructed to throw them straight back into the water without opening them and in cases of inadvertent contamination, to go straight to hospital. All the contaminated areas of the boat must be thoroughly cleaned and the fishermen's clothes and personal effects must be eliminated. Mustard gas can impregnate clothes and leather objects and persist for a long time. In fact, we have also observed cases of contamination of members of the family due to contact with the fisherman's clothing.

References

1. Hicks JH (1977) Swimming and the skin. Cutis 19:448
2. Kennedy CTC (1992) Skin hazards of swimming and diving. In: Champion RH, Burton JL, Ebling FJG (eds) Textbook of dermatology. 5th edn. Blackwell Scientific Publications, Oxford, 806
3. Chlaamidas SL Charles CR (1971) Aquagenic urticaria. Arch Dermatol 104:541
4. Davis RS, Remigio LK, Schocket AL et al (1981) Evaluation of a patient with both aquagenic and cholinergic urticaria. J Allergy Clin Immunol 68:479
5. Sibbald RG, Black AK, Eady RAJ et al (1981) Aquagenic urticaria: evidence of cholinergic and histaminergic basis. Br J Dermatol 105:297
6. Czarnetzki BM, Breethold K, Traupe H (1986) Evidence that water acts as a carrier for an epidermal antigen in aquagenic urticaria. J Am Acad Dermatol 15:623
7. Angelini G, Vena GA, Fiordalisi F (1986) Orticaria e altre dermatosi istamino-correlate. Gruppo Lepetit, Milano
8. Santoianni P, Balato N (1991) Orticarie fisiche. In: Meneghini CL, Valsecchi R, De Costanza F (eds). Orticaria angioedema. ISED, Brescia, 67
9. Cassano N, D'Argento V (1999) Orticaria. In: Angelini G, Vena GA (eds) Dermatologia professionale e ambientale. Vol III. ISED, Brescia, 875
10. Grabbe J (1998) Aquagenic urticaria. In: Henz BM, Zuberbier T, Grabbe J et al (eds) Urticaria. Springer, Berlin Heidelberg New York, 111
11. Shelley WB, Rawnsley HM (1964) Aquagenic urticaria. Contact sensitivity reaction to water. JAMA 189:895
12. Martinez-Escribano JA, Quecedo E, De la Cuadra J et al (1977) Treatment of aquagenic urticaria with PUVA and astemizole. J Am Acad Dermatol 36:118
13. Czarnetzki BM (1986) Urticaria. Springer, Berlin Heidelberg New York, 55
14. Petit A, Schnitzler L, Lassoued K et al (1992) Anti-laminin-B autoantibodies in a patient with cold urticaria. Dermatology 185:143
15. Möller A, Henning M, Zuberbier T et al (1996) Epidemiologie und Klinik der Kälteurtikaria. Hautarzt 47:510
16. Weinstock G, Arbeit L, Kaplan AP (1986) Release of prostaglandin D2 and kinins in cold urticaria and cholinergic urticaria. J Allergy Clin Immunol 77:188
17. Maltby NH, Ind PW, Causon RC et al (1989) Leukotriene E4 release in cold urticaria. Clin Exp Allergy 19:33
18. Tillie-Leblond I, Gosset P, Janin A et al (1994) Tumor necrosis factor-alpha release during systemic reaction in cold urticaria. J Allergy Clin Immunol 93:501
19. Winkelmann RH (1985) Immunofluorescent and histologic study of cold urticaria. Arch Dermatol Res 278:37
20. Neittaanmki H, Myohanen T, Fraki JE (1984) Comparison of cinnarizine, cyproheptadine, doxepin, and hydroxyzine in the treatment of idiopathic cold urticaria: usefulness of doxepin. J Am Acad Dermatol 11:483
21. Ormerod AD, Smart L, Reid TM et al (1993) Familial cold urticaria: investigation of a family and response to stanazolol. Arch Dermatol 129:343

22. Steinman HK, Greaves MW (1985) Aquagenic pruritus. J Am Acad Dermatol 13:91
23. Grandolfo M (1997) Dermatosi da agenti fisici. In: Angelini G, Vena GA (eds) Dermatologia professionale e ambientale. Vol I. ISED, Brescia, 55
24. Lotti T, Steinman HK, Greaves MW et al (1986) Increased cutaneous fibrinolytic activity in aquagenic pruritus. Int J Dermatol 25:508
25. Gola M, Francalanci S, Giorgini S et al (1986) Due casi di eczema da contatto da maschera di gomma. Boll Dermatol Allergol Profes 1:5
26. Foussereau J, Tomb R, Cavelier C (1987) Contact dermatitis to diving equipment. Boll Dermatol Allergol Profes 2:127
27. Angelini G, Vena GA (1999) Dermatologia professionale e ambientale. Voll II, III. ISED, Brescia
28. Bonamonte D (1997) Sport e dermatologia. In: Angelini G, Vena GA (eds) Dermatologia professionale e ambientale. Vol I. ISED, Brescia, 253
29. Chang W, Pien F (1988) Le infezioni contratte in ambiente marino. Stampa Medica 32:14
30. Di Napoli PL (1988) Le malattie professionali del subacqueo. Stampasma 5:12
31. Malpieri M (1988) Medicina subacquea. Leadership Medica 4:12
32. Angelini G, Vena GA, Foti C et al (1990) Dermatite da contatto con gas iprite. Boll Dermatol Allergol Profes 5:71
33. Vena GA, Foti C, Grandolfo M et al (1994) Contact irritation associated with airborne contact irritation from mustard gas. Contact Dermatitis 31:130
34. Angelini G, Vena GA (1997) Dermatosi aerotrasmesse. In: Angelini G, Vena GA (eds) Dermatologia professionale e ambientale. Vol I. ISED, Brescia, 107
35. Sartorelli E, Giubileo M, Bartolini E (1957) Contributo allo studio della bronchite cronica asmatiforme con enfisema polmonare quale postumo di intossicazione professionale da iprite. Med Lav 48:336
36. Gaffuri E, Felisi A (1957) Patologia polmonare cronica professionale da iprite. Med Lav 48:539
37. Momeni AZ, Enshaeih S, Meghdadi M et al (1992) Skin manifestations of mustard gas. Arch Dermatol 128:775
38. Gohlke H, Ullerich K (1951) Haut- und Augenschäden durch Dichlordiäthylsulfid. Hautarzt 2:404
39. Mongelli-Sciannameo N (1960) Infortunio collettivo da solfuro di etile biclorurato in un gruppo di pescatori. Rass Med Industr Igiene Lav 29:441
40. Hjorth N (1953) Food poisoning from code-rose contaminated by mustard gas. A report with 5 case histories. Acta Med Scandinav 147:237

Geographical Index

Subject Index